T0135566

Causality-Violating Lorentzian Manifolds Admitting a Shear-free Timelike Flow

vorgelegt von

Diplom-Physiker

Matthias Plaue

aus Heidelberg

Von der Fakultät II – Mathematik und Naturwissenschaften
der Technischen Universität Berlin
zur Erlangung des akademischen Grades
Doktor der Naturwissenschaften

Dr. rer. nat.

angenommene Dissertation

Promotionsausschuss:
Vorsitzender: Prof. Dr. Fredi Tröltzsch
Berichter/Gutachter: Prof. Graham S. Hall
Prof. Dr. Mike Scherfner
Prof. Dr. John M. Sullivan

Tag der wissenschaftlichen Aussprache: 16. Mai 2012

Berlin 2012

D 83

Zugl.: Diss., Technische Universität Berlin, 2012

Bibliografische Information der Deutschen Nationalbibliothek

Die Deutsche Nationalbibliothek verzeichnet diese Publikation in der
Deutschen Nationalbibliografie; detaillierte bibliografische Daten sind
im Internet über http://dnb.d-nb.de abrufbar.

©Copyright Logos Verlag Berlin GmbH 2012
Alle Rechte vorbehalten.

ISBN 978-3-8325-3186-7

Logos Verlag Berlin GmbH
Comeniushof, Gubener Str. 47,
10243 Berlin
Tel.: +49 (0)30 42 85 10 90
Fax: +49 (0)30 42 85 10 92
INTERNET: http://www.logos-verlag.de

Fakultät für Mathematik und Naturwissenschaften
der Technischen Universität Berlin

Zusammenfassung

Kausalitätsverletzende Lorentz-Mannigfaltigkeiten mit scherungsfreiem zeitartigen Fluss

von Matthias Plaue

Die vorliegende Arbeit befasst sich mit Lorentz'schen Mannigfaltigkeiten, auf welchen ein scherungsfreies, zeitartiges Vektorfeld definiert ist. Ein solches Vektorfeld ist beispielsweise dadurch charakterisiert, dass der Fluss eine Einparameter-Untergruppe konformer Transformationen der transversalen Metrik darstellt.

Eine weitere Charakterisierung ist durch die Tatsache gegeben, dass die Metrik – zumindest lokal – als "gekipptes" Produkt dargestellt werden kann, welches der Metrik von standardstationären Raumzeiten ähnelt:

$$g(t,x) = -(\mathrm{d}t - b(t,x))^2 + a(t,x)^2 h(x).$$

Im Gegensatz zu standardstationären Raumzeiten wird keine stabile Kausalität gefordert, und einige Resultate zur Kausalstruktur werden präsentiert. Diese Resultate basieren zum Teil auf der Eichfreiheit $b \mapsto b + \mathrm{d}f$ sowie gewisser Sobolev-Ungleichungen für Differenzialformen.

Darüber hinaus werden für den Fall $a(t,x) = a(t)$ Killing-Symmetrien diskutiert und die Ricci-Krümmung berechnet. Beispiele von Lösungen der Einstein-Hilbert-Gleichung werden konstruiert, bei denen der scherungsfreie Fluss ein expandierendes und rotierendes Fluid mit nicht-verschwindendem Wärmestrom darstellt. Der Fluss kann geodätisch oder konform stationär sein, und die Raumzeit kann, in Abhängigkeit von einem Parameter, geschlossene zeitartige Kurven enthalten. Außerdem wird eine Charakterisierung von Lösungen mit geodätischem und barotropem idealen Fluid oder Staubmaterie geliefert.

Schließlich werden Eigenschaften von Raumzeiten vom Gödel-Typ gezeigt, die sich als Lösungen mit geladenem idealem Fluid interpretieren lassen.

School for Mathematics and Natural Sciences
of Technische Universität Berlin

Abstract

Causality-Violating Lorentzian Manifolds Admitting a Shear-free Timelike Flow

by Matthias Plaue

We investigate Lorentzian manifolds that admit a shear-free timelike vector field. Such a vector field is characterized, for example, by the fact that its flow is a one-parameter group of conformal transformations with respect to the transversal metric.

Another characterization is given by the fact that the metric may—locally, at least—be written in a "tilted" product form similar to standard stationary spacetimes:

$$g(t, x) = -(\mathrm{d}t - b(t, x))^2 + a(t, x)^2 h(x).$$

However, in contrast to standard stationary spacetimes, we do not require the spacetime to be stably causal, and we show some results concerning the causal structure. Some of these results are based on the gauge freedom $b \mapsto b + \mathrm{d}f$ and certain Sobolev inequalities for differential forms.

Furthermore, we discuss Killing symmetries and compute the Ricci curvature of tilted products with $a(t, x) = a(t)$. Examples of solutions to the Einstein–Hilbert equation are given where the shear-free flow represents an expanding and rotating viscous fluid with non-vanishing heat flow. We find geodesic and conformally stationary solutions which may contain closed timelike curves, depending on a certain parameter. We also give a characterization of geodesic barotropic perfect fluid and dust solutions.

Finally, we take note of some properties of charged perfect fluid solutions of Gödel type.

TABLE OF CONTENTS

Page

i

ACKNOWLEDGMENTS

It is a pleasure to thank the many people who have been providing me with moral, professional and financial support during the making of this thesis.

First and foremost I would like to express my sincere gratitude to my supervisor, M. Scherfner, whose fierce commitment and generous spirit are always an inspiration, and whose academic and human experience has proved to be invaluable to me.

Furthermore, I would like to acknowledge Technische Universität Berlin, and also the University of Heidelberg. Very special appreciation goes out to my employers without whom it would probably not have been possible for me to pursue my academic interests. Besides M. Scherfner, those have been G. Bärwolff, D. Ferus, F. A. Hamprecht, H. Schwandt, and E. Zorn. It has been a pleasant obligation to work in fields as diverse as image processing, human crowd analysis, and mathematics education for engineers, and special thanks go out to my colleagues who have walked those roads with me.

I also wish to express my gratitude to G. S. Hall and J. M. Sullivan for reviewing my thesis, and F. Tröltzsch for agreeing to officiate as chairman.

Furthermore, I am greatly indebted to M. Gürses for his kind invitation to visit Bilkent University. Many results contained in this thesis are based on those most productive dicussions. I also owe many thanks to the organizers of the Lorentzian geometry meeting in Granada 2011, and M. Ortega in particular, for their funding that made my visit to this wonderful conference possible.

Special acknowledgment goes out to S. Born and A. Dirmeier for many inspiring discussions, and to A. Dirmeier, once more, for his proofreading of this thesis, although all errors are my own.

Additionally, I would like to mention H.-H. von Borzeszkowski, T. Chrobok, and K.-E. Hellwig who not only supervised my diploma thesis but have been organizing a stimulating seminar on geometrical methods in mathematical physics at the TU Berlin for quite some time now.

I also wish to thank all the other kind people I have met, if only briefly, who have added to my graduate experience through fruitful discussions, or in any other way.

Finally, I especially thank my family and friends for their sympathy and encouragement.

DEDICATION

To my beloved parents who taught me to never back down.

Chapter 1

INTRODUCTION

An important subject in the study of the causal structure of (time-oriented) Lorentz-ian manifolds (i.e., spacetimes) are curves of a fixed causal character, where timelike or non-spacelike curves are of particular interest. The conformal geometry of the spacetime plays a prominent role in this context since the causal character is a conformal invariant. It should be noted, however, that the classification of conformal and causal structures may not be seen as an entirely equivalent problem [GPS03].

Examples of spacetimes that—depending on context—exhibit an "interesting" or "pathological" causal structure are those that contain closed timelike curves. Such spacetimes are called non-chronological. Apart from being chronological, Lorentzian manifolds may satisfy a number of other causality conditions which are shown in Table 1.1, cf. [BEE96, MS06]. Globally hyperbolic spacetimes are the models with the "nicest" causal structure and may be understood as the Lorentzian analogue to Riemannian complete manifolds. For example, Minkowski space and Friedmann–Lemaître–Robertson–Walker (FLRW) models (with complete fiber) are globally hyperbolic.

Although not a conformal invariant, spacetime singularities are also sometimes regarded as part of the causal structure. Furthermore, the celebrated Penrose–Hawking singularity theorems [HE73] hold explicitly for spacetimes which are chronological, although this requirement may be lessened [Kri90].

An elementary example for a non-chronological spacetime is the cylinder $S^1 \times \mathbb{R}$ with metric $g = -\mathrm{d}\phi^2 + \mathrm{d}x^2$ (*Lorentz cylinder*) – curves with constant x-coordinate are closed and timelike. Important spacetimes in the middle of the causal hierarchy are, for example, the plane fronted gravitational waves: these are not causally simple but causally continuous [FS03].

As can be seen from the Lorentz cylinder, examples of non-chronological spacetimes can be constructed very easily. However, this work is mostly concerned with causality violation that is "non-trivial" in the sense of Carter [Car68], i.e., closed timelike or non-spacelike curves which are homotopically equivalent to a point and may therefore not be "unwrapped" by passing to the universal Lorentzian covering space. Thus, the causality violation that is exhibited by the Lorentz cylinder is trivial – the closed time-like curves contained in the famous Gödel spacetime [Göd49], which is homeomorphic

```
globally hyperbolic
        ⇓
  causally simple
        ⇓
causally continuous
        ⇓
   stably causal
        ⇓
  strongly causal
        ⇓
  distinguishing
        ⇓
      causal
        ⇓
  chronological
        ⇓
non-totally vicious
```

Table 1.1: Causality conditions

to $\mathbb{R}^4$, are not. Since Gödel's seminal work, the Gödel metric has been generalized or modified by a number of authors [Pan83, RT83, Aga84, VBS84, Obu90, Obu92, KO93, Obu00, BP03]. The main motivation of this work is the construction and analysis of models exhibiting an interesting causal structure which are otherwise "unsuspicious", i.e., simply connected and non-compact. Nevertheless, note that we also investigate spacetimes containing a compact timelike submanifold.

Furthermore, we are not only concerned with the geometric but also the physical aspects of such models: spacetimes containing closed timelike curves may be interpreted as models where an observer might be able to visit an event in his own past. The physical reality of such models is highly hypothetical. In fact, it is rejected by many physicists, most prominently by Hawking who argues that some mechanism must prevent the emergence of closed timelike curves on macroscopic scales (*Chronology Protection Conjecture* or *Chronology Protection Axiom* [Haw92]). However, we would like to contrast with a quote from another expert in the field, Visser [Vis97, p. 266]:

> [...] Adopting this point of view implies that the status of chronology protection has been elevated to an empirical experimental question. The chronology protection axiom can be overturned only by the empirical con-

struction of a time machine – a task that we are unlikely to accomplish in the immediate future.

Examples of non-chronological spacetimes which are physically motivated include the van Stockum dust [Lan23, vS37], the above mentioned Gödel cosmos [Göd49], the Wheeler worm hole [FW62], the Kerr vacuum [Ker63], the Tipler cylinder [Tip74], the Gott time machine [GI91], the Ori–Soen time machine [SO96], and the Krasnikov tube [ER97]. However, by the writing of this thesis and to the author's best knowledge no model exists that may be considered as fully physically valid concerning requirements such as energy conditions, stability, existence of a partial Cauchy surface etc. Other objections to some of the models include that no information may exit the causality violating region, or that the model represents a cosmological solution that simply does not fit our own observable universe.

While there is good reason to believe that it is impossible to *build* a time machine in the context of classical general relativity (see, for example, [Kra02]), these arguments are not necessarily applicable to cosmological models, and the possibility of closed timelike curves in an early epoch of the universe cannot be easily dismissed; see [GIL98, Car00], or the more speculative essay [Min09a]. This situation may perhaps be compared to the standing of the Cosmic Censorship Hypothesis: the formation of a naked singularity may not be allowed by the laws of nature – although according to the cosmological standard model the Big Bang clearly is such a singularity in an early epoch of the universe.

Formally, a cosmological model may be characterized as a Lorentzian manifold endowed with a distinguished timelike vector field (*observer field*) the integral curves of which represent the world lines of the galactic matter. The projection of the covariant derivative of this vector field onto its restspace may be decomposed into irreducible parts: the vorticity, expansion and shear. The terminology describes precisely what each component "does", namely distort small spatial volumes as indicated in Figure 1.1. Furthermore, the component of the covariant derivative parallel to the observer field, the acceleration, vanishes if and only if the integral curves are geodesics, i.e., if the galactic matter is in free fall and subject to no other forces. The evolution of these *kinematical quantities* along the galactic observers' world lines is determined by a set of identities called Raychaudhuri's equations [Ray55, KS07], where the equation describing the expansion is the most prominent and usually referred to as *the* Raychaudhuri equation. The latter is of great importance in establishing the singularity theorems [HE73]. Although the dynamics therefore seem to be explicitly known, there are still important open problems concerning the kinematics of spacetimes. One of the most prominent conjectures is perhaps the statement that shear-free barotropic perfect fluids cannot both be rotating and expanding (*Shear-free Fluid Conjecture*). See, for example, [SSS98] for a more recent discussion of the problem.

In this work, we pay special attention to shear-free kinematical manifolds, which appear frequently in general relativity. For example, any (conformally) stationary, rigidly rotating or first-order Hubble-isotropic [HP99] model is necessarily shear-free.

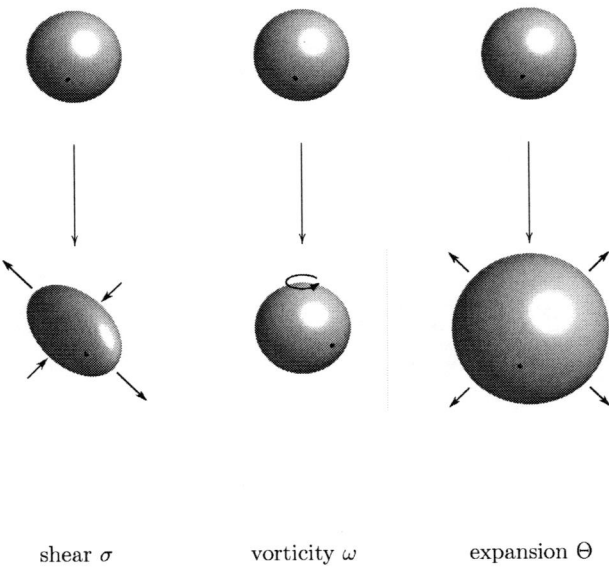

shear σ vorticity ω expansion Θ

Figure 1.1: The kinematical quantities and how they contribute to changes of small volumes in restspace with respect to the proper time of the observer.

1.1 Motivation and Aim

It has recently been observed that given a cosmological model admitting a shear-free reference frame, one may take advantage of the local coordinate expression of the metric adapted to this reference frame to construct models with different kinematical properties [COS01]. Moreover, it was observed that this technique may be applied to totally vicious spacetimes (for example, the Gödel spacetime) to yield expanding and rotating models that admit causally well-behaved regions [Sch00, GPS+10]. In this context, our work mainly deals with two important issues:

- Firstly, the construction is based on local coordinate expressions. However, causality theory of Lorentzian manifolds is inherently global, since any point of a spacetime is contained in a causally well-behaved neighborhood. The argument typically used in the physics literature showing the absence of closed timelike curves is based on the fact that some (globally defined) coordinate function is a temporal function, see Remark 2.3.20 below.

- Secondly, the construction generally does not yield a physically meaningful solution to the Einstein–Hilbert equation.

After introducing some basic notions in Chapter 2, we discuss manifolds admitting reference frames in Chapter 3. Reference frames may be understood as unit cospacelike vector fields, and are therefore not only defined for spacetimes but for Riemannian manifolds as well. Mathematically, a reference frame induces a particular almost product structure, i.e., a splitting of the spacetime's tangent bundle [Gra67]. Physically, projecting spacetime tensors onto the horizontal distribution, i.e., the restspace, yields the Newtonian quantities measured by observers travelling in this reference frame ("$(D + 1)$-decomposition"). Consequently, a reference frame is also called an observer field. The decomposition of the observer field's covariant derivative yields the kinematics of the model. We identify the conformally invariant kinematical quantities, present an exhaustive characterization of shear-free and conformally stationary reference frames, and characterize stably causal conformally stationary spacetimes in terms of kinematics.

In Chapter 4, based on the local characterization of shear-free kinematical manifolds, we introduce a global product structure that we term *tilted twisted products*. We investigate the causal structure and present particular cases when Killing vector fields defined on the fiber are inherited by the total space. Also, we compute the Ricci curvature for tilted warped products, i.e., when the twisting function only depends on the standard observers' proper time.

Examples of tilted product spacetimes that are solutions to the Einstein–Hilbert equation for an isotropic fluid with non-vanishing heat flow are presented in Chapter 5. Furthermore, we classify the tilted warped products which are geodesic barotropic perfect fluid solutions; these are either of FLRW or of Gödel type.

Chapter 6 is dedicated to the application of global techniques to stationary spacetimes of tilted product type. In the first section of this chapter, we state some results on relations between the gauge freedom of changing the slicing and stable causality. The second section deals with some properties of charged perfect fluid solutions of Gödel type. Finally, in Chapter 7 we give an outlook on further opportunities of research.

6

Some results of this thesis have previously been published in [DPS09], [GPS+10] and [GPS11]. To verify results, and generate the figures, the computer algebra system Maple [Map] (with the package GRTensorII [MPL96]) and the numerical computing language MATLAB [MAT] were used.

Chapter 2

PRELIMINARIES

Mathematically, this work is mainly concerned with structures on *manifolds*. By a manifold M, we mean a smooth manifold without boundary, i.e., a second countable Hausdorff topological space, locally homeomorphic to $\mathbb{R}^d$ (where $d = \dim(M) \in \mathbb{N}$ is fixed), endowed with an equivalence class of atlases the charts of which overlap smoothly. The ring of real-valued (complex-valued) smooth functions on M is denoted by $\mathfrak{F}(M) = \mathfrak{F}^{\mathbb{R}}(M)$ ($\mathfrak{F}^{\mathbb{C}}(M)$). If the differentiability class of a map is not explicitly stated, it is always assumed to be smooth.

For further reading on the mathematical topics that are relevant for this work we may refer to a number of text books, such as, for example, [CBDMDB82, O'N83, Bes87, BEE96, KN96, Con01, Ish01, Jos02, Ber03].

2.1 Basic Manifold Theory

In the following, we recall some basic notions of manifold theory (without a metric) to fix notation. In this section, let M, N be smooth manifolds and $d = \dim(M)$.

2.1.1 Tangent Spaces and Tensor Fields.

Definition 2.1.1. Let $p \in M$. A *tangent vector* v at p is a map $v \colon \mathfrak{F}(M) \to \mathbb{R}$ with the following properties:

1. The map v ist linear; for all $f, g \in \mathfrak{F}(M)$ and $\lambda \in \mathbb{R}$ we have:
$$v(f + g) = v(f) + v(g), \quad v(\lambda f) = \lambda v(f).$$

2. For all $f, g \in \mathfrak{F}(M)$, *Leibniz' product rule* holds:
$$v(f \cdot g) = v(f) \cdot g(p) + f(p) \cdot v(g).$$

Example 2.1.2. Consider a connected non-empty open subset of the real line (i.e., an open interval) $I \subset \mathbb{R}$ as a one-dimensional manifold. For each $s \in I$, the map $\partial_t|_s \colon \mathfrak{F}(I) \to \mathbb{R}$, $\partial_t|_s f = f'(s)$ is a tangent vector at s, where f' is the derivative of f.

Theorem 2.1.3. *The set of tangent vectors at a given point $p \in M$, together with the usual addition and multiplication with scalars, is a vector space of dimension d, the* tangent space T_pM. *Given natural coordinates $(x_0, \ldots, x_{d-1})$ on $\mathbb{R}^d$, for a chart (U, ϕ) of M with $p \in U$, the maps*

$$\partial_i|_p \colon \mathfrak{F}(M) \to \mathbb{R}, \; \partial_i|_p f = \frac{\partial(f \circ \phi^{-1})}{\partial x_i}(\phi(p)), \quad i = 0, \ldots, d-1,$$

constitute a basis of T_pM, the coordinate basis *at p with respect to (U, ϕ).*

Definition 2.1.4. We denote the dual space of T_pM by T_pM^*.

1. A (contravariant) *vector field* X is a map assigning to each $p \in M$ a tangent vector $X_p \in T_pM$ that is smooth in the sense that the function $X(f) \colon p \mapsto X_p(f)$ is smooth for any $f \in \mathfrak{F}(M)$.

2. A covariant vector field or *one-form* η is a map assigning to each $p \in M$ a covector $\eta_p \in T_pM^*$ such that the function $\eta(X) \colon p \mapsto \eta_p(X_p)$ is smooth for any vector field X.

3. In general, a (smooth) (r, s)-tensor field τ is a map assigning to each $p \in M$ a tensor $\tau_p \in \underbrace{T_pM \otimes \cdots \otimes T_pM}_{r \text{ times}} \otimes \underbrace{T_pM^* \otimes \cdots \otimes T_pM^*}_{s \text{ times}}$ such that the function

$$\tau(\eta_1, \ldots, \eta_r, X_1, \ldots, X_s) \colon p \mapsto \tau_p(\eta_1|_p, \ldots, \eta_r|_p, X_1|_p, \ldots, X_s|_p)$$

is smooth for any one-forms $\eta_1, \ldots, \eta_r$ and vector fields $X_1, \ldots, X_s$.

Any tensor field, or similarly, any tensor at a point, is locally characterized on a chart by its components with respect to a given *coordinate frame* $(\partial_0, \ldots, \partial_{d-1})$ and its dual basis $(\mathrm{d}x^0, \ldots, \mathrm{d}x^{d-1})$:

$$\tau = \tau^{j_1 \ldots j_r}{}_{i_1 \ldots i_s} \partial_{j_1} \otimes \cdots \otimes \partial_{j_r} \otimes \mathrm{d}x^{i_1} \otimes \cdots \otimes \mathrm{d}x^{i_s}$$

Here and throughout this work, we adopt Einstein's summation convention, i.e., summation over repeated indices is understood. A tensor field is smooth if and only if its components with respect to some (thus any) local basis are smooth functions.

Furthermore, we denote by $\mathrm{sym}(\tau)$ and $\mathrm{asym}(\tau)$ the symmetric and skew-symmetric parts of a covariant $(0, s)$-tensor τ, normalized such that $\tau = \mathrm{sym}(\tau) + \mathrm{asym}(\tau)$. For the skew-symmetric and symmetric tensor product we write $\tau \wedge \mu$ and $\tau \vee \mu$, respectively. We choose these products not to be normalized – for example, $u \vee v = u \otimes v + v \otimes u$ for one-forms u, v. In local coordinates, (anti-)symmetrization is denoted by (brackets) parantheses, for example

$$(\mathrm{sym}(u \otimes v))_{ij} = \frac{1}{2}(u \vee v)_{ij} = u_{(i}v_{j)},$$

$$(\mathrm{asym}(u \otimes v))_{ij} = \frac{1}{2}(u \wedge v)_{ij} = u_{[i}v_{j]}.$$

2.1.2 Differentials, Lie Derivatives and Flows. Submanifolds.

Definition 2.1.5. Let $f \colon M \to N$ be a map. The *differential* of f at $p \in M$ is defined as the linear map $f_{*p} \colon T_p M \to T_{f(p)} N$, given for arbitrary $g \in \mathfrak{F}(N)$ by

$$(f_{*p} v)(g) = v(g \circ f),$$

where "$\circ$" denotes the composition of maps. We also say that $f_{*p} v$ is the *push-forward* of v via f.

The *pull-back* of a covariant $(0, s)$-tensor field τ on N via f is the tensor field $f^* \tau$ on M, defined by

$$(f^* \tau)_p(v_1, \ldots, v_s) = \tau_{f(p)}(f_{*p} v_1, \ldots, f_{*p} v_s)$$

for each $p \in M$ and any $v_1, \ldots, v_s \in T_p M$.

Remark 2.1.6. For $\tau \in \mathfrak{F}(M)$, the pull-back is simply given by $f^* \tau = \tau \circ f$.

Definition 2.1.7. Let X be a vector field on M. An *integral curve* of X through $q \in M$ is a path $\gamma^q \colon I_q \to M$ ($0 \in I_q \subset \mathbb{R}$ an open interval) such that

$$\gamma^q_*(\partial_t|_s) = X_{\gamma(s)} \text{ and } \gamma^q(0) = q$$

hold for each $s \in I_q$.

For each point $p \in M$, there exists an open interval I and a neighborhood U of p such that we may define a map $\psi \colon I \times U \to M$ by

$$\psi(t, p) = \gamma^p(t).$$

This map is called the *local flow* of X and we write $\psi_t(p) := \psi(t, p)$.

Proposition 2.1.8. *The local flow $\psi_t \colon U \to M$ is a diffeomorphism onto its image for each $t \in I$. Furthermore, we have the following:*

1. *The map ψ_0 is the identity map of U.*

2. *It holds that $\psi_{t+s} = \psi_s \circ \psi_t$ whenever $s, t, s + t \in I$.*

Definition 2.1.9. Let τ be a covariant tensor field on M. The *Lie derivative* of τ with respect to a vector field X on M is given by

$$\mathcal{L}_X \tau = \lim_{t \to 0} \frac{1}{t}(\psi_t^* \tau - \tau),$$

where ψ_t is the local flow of X.

Definition 2.1.10. Let X and Y be vector fields on M. The *Lie bracket* of X and Y is the vector field $[X,Y]$ given by

$$[X,Y](f) = (X \circ Y)(f) - (Y \circ X)(f)$$

for all $f \in \mathfrak{F}(M)$.

Remark 2.1.11. In local coordinates,

$$[X,Y]_i = X^k \partial_k Y_i - Y^k \partial_k X_i.$$

By applying rules for tensor derivation (see Definition 2.2.12 below), the Lie derivative may be defined for tensor fields of arbitrary rank. It turns out that the Lie bracket is in fact a Lie derivative: $\mathcal{L}_X Y = [X,Y]$.

Definition 2.1.12. A topological subspace S of M is called a *submanifold* of M if for some fixed $k \in \mathbb{N}$, about any point $p \in S$ there exists a chart (U,ϕ) such that $\phi(U \cap S) = \mathbb{R}^k \cap \phi(U)$, where we understand $\mathbb{R}^k$ as a subspace of $\mathbb{R}^d$ in the canonical way. Such a chart or coordinates are called *adapted* to S.

We have that S is a k-dimensional manifold with an atlas given by $(U_i \cap S, \phi_i|_{U_i \cap S})_{i \in \mathfrak{I}}$, where $(U,\phi)_{i \in \mathfrak{I}}$ is a collection of adapted charts covering S. The inclusion map $i \colon S \to M$ is smooth and i_{*p} is injective for any $p \in S$. The tangent space $T_p S$ of S is isomorphic to the image of i_{*p} and may therefore be regarded as a subspace of $T_p M$. A *hypersurface* is a submanifold of dimension $k = d - 1$.

Example 2.1.13. Let $M = P \times Q$, where P and Q are manifolds. M can naturally be endowed with an atlas by taking products of charts of P and Q. For each $(p,q) \in M$, $P_q := P \times \{q\}$ and $Q_p := \{p\} \times Q$ are submanifolds of M, and $T_{(p,q)}M \cong T_{(p,q)}P_q \oplus T_{(p,q)}Q_p \cong T_p P \oplus T_q Q$.

Definition 2.1.14. Let $M = P \times Q$ be a product manifold, and $\pi_1 \colon M \to P$, $\pi_2 \colon M \to Q$ be the projections.

- The *lift* $\hat{\tau}$ of a covariant tensor field τ defined on either P or Q is the pull-back of τ with respect to the projection: $\hat{\tau} = \pi_1^* \tau$, or $\hat{\tau} = \pi_2^* \tau$.

- The lift of a contravariant $(r,0)$-tensor field τ on P is defined as the unique tensor field $\hat{\tau}$ on M with $\hat{\tau}(\hat{\eta}^1, \ldots, \hat{\eta}^r) = \tau(\eta^1, \ldots, \eta^r)$ for all one-forms $\eta^1, \ldots, \eta^r$ on P and $\hat{\tau}(\hat{\eta}^1, \ldots, \hat{\eta}^r) = 0$ for all one-forms $\eta^1, \ldots, \eta^r$ on Q. Lifts from Q are defined dually.

Remark 2.1.15. Let $(\partial_0, \ldots, \partial_{d-1})$ be a frame adapted to the splitting $M = P \times Q$, so that $T_p P$ is spanned by $(\partial_0, \ldots, \partial_{k-1})$ where $k = \dim(P)$. Given a vector field $X = X^i \partial_i$ on P, the coordinates of its lift $\hat{X}$ are simply given by $\hat{X}^i = X^i$ for $i = 0, \ldots, k-1$ and $\hat{X}^i = 0$ for $i = k, \ldots, d-1$. Similarly, lifting a one-form $\eta = \eta_i \mathrm{d}x^i$ on P means setting all coordinate functions $\eta_k, \ldots, \eta_{d-1}$ to zero.

2.2 Vector Bundles, Metrics and Connections

In this section, let M, N be manifolds and $d = \dim(M)$. Throughout, let either $\mathbb{K} = \mathbb{R}$ or $\mathbb{K} = \mathbb{C}$ be the field of real or complex numbers. Informally, a property defined for M and mappings $f, g, \ldots$ from M into some other sets is said to hold *locally* if for each $p \in M$ there exists an open neighborhood U of p such that the property holds for U and the restrictions of $f, g, \ldots$ to U.

2.2.1 Vector Bundles.

Definition 2.2.1. Let $k \in \mathbb{N}$. A (smooth) $\mathbb{K}$-*vector bundle* of rank k over the *base manifold* M is a manifold E together with a surjective map $\pi \colon E \to M$ (the *projection*) such that:

1. For every $p \in M$, $E_p := \pi^{-1}(\{p\})$ carries the structure of a $\mathbb{K}$-vector space,

2. there exists a covering $(U_i)_{i \in \mathfrak{I}}$ of M, such that for each index $i \in \mathfrak{I}$, there exists a diffeomorphism $\phi_i \colon U_i \times \mathbb{K}^k \to \pi^{-1}(U_i)$, for which we have the following:

   - for each $v \in \mathbb{K}^k$, the map $(\pi \circ \phi_i)(\,\cdot\,, v)$ is the identity on U_i,
   - for each $p \in U_i$, the map $\phi_i(p, \,\cdot\,)$ is a linear isomorphism of $\mathbb{K}^k$ onto E_p.

The collection $(U_i, \phi_i)_{i \in \mathfrak{I}}$ is called a *local trivialization* of E, E_p is called the *fiber* at p.

A vector bundle may be seen as a collection of vector spaces parameterized by M such that every point $p \in M$ is assigned to a vector space E_p that is isomorphic to $\mathbb{K}^k$. A local trivialization guarantees that this assignment is smooth.

Example 2.2.2. The disjoint union of all tangent spaces,

$$TM := \bigsqcup_{p \in M} T_p M = \bigcup_{p \in M} (\{p\} \times T_p M),$$

with natural projection $\pi \colon TM \to M$, $\pi(p, v) = p$, is a real vector bundle over M of rank d, called the *tangent bundle* of M: Given a chart (U, ϕ) on M, let $(\partial_l)_{l=0,\ldots,d-1}$ be the coordinate basis, and for each $p \in U$ let $K_p \colon T_p M \to \mathbb{R}^d$ be the corresponding coordinate map on $T_p M$. The collection of $(TU, T\phi)$'s with $T\phi(p, w) = (\phi(p), K_p(w))$ defines an atlas for TM, making it a $2d$-dimensional manifold. Define $\psi \colon U \times \mathbb{R}^d \to \pi^{-1}(U)$, $\psi(p, v) := (p, K_p^{-1}(v))$; then, the collection of all such (U, ψ)'s defines a local trivialization of TM.

Definition 2.2.3. Let E and F be two $\mathbb{K}$-vector bundles over M of rank k and l, respectively. We may define the *direct sum* $E \oplus F$ of these two bundles by setting

$E \oplus F := \bigsqcup_{p \in M}(E_p \oplus F_p)$ with the natural projection $\pi \colon E \oplus F \to M$ mapping each vector in some fiber $E_p \oplus F_p$ to its base point p. Vectors in E_p or F_p may be identified as vectors in $E_p \oplus F_p$ in the usual way, and any $v \in E_p \oplus F_p$ may be uniquely decomposed as a sum $v = v_E + v_F$ with $v_E \in E_p$, $v_F \in F_p$. Given local trivializations $(V_i, \psi_i)_{i \in \mathfrak{I}}$ and $(W_m, \chi_m)_{m \in \mathfrak{M}}$ of E and F respectively, we have a local trivialization of $E \oplus F$ via the collection of all $U \subset M$ with $U = V_i \cap W_m \neq \emptyset$ and the maps $\phi \colon U \times \mathbb{K}^{k+l} \to \pi^{-1}(U)$, $\phi(p, (e, f)) := \psi_i(p, e) + \chi_m(p, f)$ where $e \in \mathbb{K}^k$, $f \in \mathbb{K}^l$.

Similarly, through fiberwise operations, we define the (symmetric, skew-symmetric) *tensor product*, *dual bundle* and *complex conjugate bundle*: $E \vee F$, $E \wedge F$, $E \otimes F$, E^*, $\bar{E}$.

Definition 2.2.4. A *section* X of a vector bundle E with projection $\pi \colon E \to M$ is a map $M \to E$ with $\pi \circ X = \mathrm{id}_M$.

A section may be viewed as a smooth assignment $M \ni p \mapsto X_p \in E_p$. The set of sections is an $\mathfrak{F}^{\mathbb{K}}(M)$-module in a natural way, which we denote with $\Gamma(E)$: For any $f \in \mathfrak{F}^{\mathbb{K}}(M)$, $X, Y \in \Gamma(E)$ define the sections $(fX)_p := f(p)X_p$ and $(X + Y)_p := X_p + Y_p$.

Example 2.2.5. The set of sections of the trivial bundle $M \times \mathbb{K}$ may be identified with $\mathfrak{F}^{\mathbb{K}}(M)$. A vector field on a manifold is an element of $\Gamma(TM)$, a one-form is an element of $\Gamma(TM^*)$, and so on.

Remark 2.2.6. If for two vector bundles $\pi_1 \colon E \to M$ and $\pi_2 \colon F \to N$ there are diffeomorphisms $f \colon E \to F$ and $g \colon M \to N$ with $\pi_2 \circ f = g \circ \pi_1$, these bundles are isomorphic. We say that E is *trivial* if it is isomorphic to a standard trivial bundle $N \times \mathbb{K}^k \to N$. A vector bundle is trivial if and only if there exist k sections that are linearly independent at each point in M.

Remark 2.2.7. A section X of a vector bundle may be characterized locally as follows: Given a chart (U, ϕ) on M and a local trivialization (V, ψ) of E, the map $\pi_2 \circ \psi^{-1} \circ X \circ \phi^{-1}$ is a smooth $\mathbb{K}^k$-valued vector field on $\phi(U \cap V) \subset \mathbb{K}^d$ in the usual sense, where $\pi_2 \colon V \times \mathbb{K}^k \to \mathbb{K}^k$ is the natural projection onto the second factor.

2.2.2 Metrics and Connections. Parallel Transport.

In the following, let E be a $\mathbb{K}$-vector bundle of rank k over some manifold M. Fiberwise vector bundle operations induce the same operations on the space of sections, i.e., there are the following isomorphisms of modules: $\underline{\Gamma(E \oplus F)} \cong \Gamma(E) \oplus \Gamma(F)$, $\Gamma(E \otimes F) \cong \Gamma(E) \otimes \Gamma(F)$, $\Gamma(E^*) \cong \Gamma(E)^*$, $\Gamma(\bar{E}) \cong \overline{\Gamma(E)}$. For example, a one-form η may equivalently be viewed as an element of $\Gamma(TM^*)$, i.e., a smooth assignment $M \ni p \mapsto \eta_p \in T_pM^*$, or as an element of $\Gamma(TM)^*$, i.e., an $\mathfrak{F}(M)$-linear map $\eta \colon \Gamma(TM) \to \mathfrak{F}(M)$. For details on the subject, we refer to [Con01].

Recall that the *index* of a bilinear form is given by the largest number that is a dimension of a subspace on which the form is negative definite. Equivalently, the index of a bilinear form is the number of negative eigenvalues of its Gramian matrix computed with respect to some basis.

Definition 2.2.8. If E is a real vector bundle, we call $g \in \Gamma(E^* \vee E^*)$ a *vector bundle metric* (over E) if g_p is a non-degenerate bilinear form over E_p for each $p \in M$, with fixed index $\iota \in \{0, \ldots, k\}$ independent of p. If $\iota = 0$, we say that g is a *Riemannian metric*; if $\iota = 1$ and $k \geq 2$, we say that g is a *Lorentzian metric*.

Any vector bundle metric g over TM is called a *semi-Riemannian metric*, and the ordered pair (M, g) a *semi-Riemannian manifold*. If g is Riemannian (Lorentzian), then (M, g) is also called a Riemannian (Lorentzian) manifold.

Remark 2.2.9. Similarly, Hermitian vector bundle metrics over complex vector bundles are defined as sections of $\bar{E}^* \vee E^*$. With the convention used here, Lorentzian metrics have signature $(-, +, \ldots, +)$.

Fix a vector bundle metric g, and $p \in M$. As it is well-known from linear algebra, the inner product $\langle \cdot, \cdot \rangle_p = g_p(\cdot, \cdot)$ may be extended to general $(r, 0)$ tensors via the rule

$$\langle v \otimes w, v \otimes w \rangle_p = \langle v, v \rangle_p \langle w, w \rangle_p$$

for $v, w \in E_p$. These inner products vary smoothly over the manifold to define vector bundle metrics over E^r.

Furthermore, via the non-degeneracy of g_p, there are the *musical isomorphisms* $E_p \to E_p^*$, $v \mapsto v^\flat := g_p(v, \cdot)$ and its inverse $\eta \mapsto \eta^\sharp$. These isomorphisms extend naturally to the tensor bundles of higher order and sections of these bundles as well.

In terms of coordinates provided by local trivializations, the musical isomorphisms correspond to the lowering and raising of indices: $(v^\flat)_i = g_{mi} v^m$, $(\eta^\sharp)^i = g^{mi} \eta_m$. The obvious formula $g_p(v, w) = w^\flat(v) = v^\flat(w)$, for example, translates to $g_{mn} v^m w^n = v^m w_m = v_m w^m$. The matrix (g^{ij}) is the inverse of (g_{ij}): $g^{im} g_{mj} = \delta^i{}_j$, where $\delta^i{}_j$ is the Kronecker symbol. Of course, for tensors of higher order there are actually a number of isomorphisms since we may choose to raise or lower any number of indices.

Finally, we may also define the *Hodge dual* with respect to $\langle \cdot, \cdot \rangle_p$: Given an orientation of E_p, for each $s = 0, \ldots, k$ it is the unique isomorphism $*_p \colon \bigwedge^s E_p \to \bigwedge^{k-s} E_p$ with

$$s! \langle \tau, \tau \rangle_p \epsilon_p = \tau \wedge *\tau,$$

where $\epsilon_p \in \bigwedge^k E_p$ is the normalized volume form, i.e., $\epsilon_p(X_1, \ldots, X_n) = 1$ for any orthonormal, positively oriented base $(X_1, \ldots, X_n)$ of E_p. We say that the bundle E is *oriented* if the transition functions between local trivializations preserve orientation. In this case, $*$ also defines an isomorphism of sections: $* \colon \Gamma(\bigwedge^s E) \to \Gamma(\bigwedge^{k-s} E)$. A manifold is called *orientable* if there exists an atlas such that its tangent bundle is oriented with respect to this atlas.

Definition 2.2.10. If g is a Riemannian or Hermitian metric, for any tensor field τ define the *(pointwise) norm* as the function

$$|\tau| = |\tau|_g = \sqrt{\langle \tau, \tau \rangle}.$$

Remark 2.2.11. The pointwise norm may be computed from a tensor's components via the formula

$$|\tau|^2 = g_{b_1 d_1} \cdots g_{b_r d_r} g^{a_1 c_1} \cdots g^{a_s c_s} \tau^{b_1 \ldots b_r}_{ a_1 \ldots a_s} \tau^{d_1 \ldots d_r}_{ c_1 \ldots c_s}.$$

Definition 2.2.12. A map

$$\Gamma(TM) \times \Gamma(E) \to \Gamma(E), (X, \psi) \mapsto \nabla_X \psi$$

is called a *connection* or *covariant derivation* over E if for each $f \in \mathfrak{F}^{\mathbb{K}}(M)$, $X \in \Gamma(TM)$ and $\psi, \phi \in \Gamma(E)$ the following holds:

1. The map $Z \mapsto \nabla_Z \psi$ is $\mathfrak{F}^{\mathbb{K}}(M)$-linear,
2. $\nabla_X(f\psi) = X(f)\nabla_X\psi + f\nabla_X\psi$,
3. $\nabla_X(\psi + \phi) = \nabla_X\psi + \nabla_X\phi$.

Given a connection ∇, we always assume that it is extended to sections of the tensor bundles $E \otimes \cdots \otimes E \otimes E^* \otimes \cdots \otimes E^*$ by linearity and the tensor derivation rules:

1. $\nabla_X f = X(f)$,
2. $\nabla_X(\psi \otimes \phi) = (\nabla_X\psi) \otimes \phi + \psi \otimes (\nabla_X\phi)$,
3. $\nabla_X(\mathcal{C}T) = \mathcal{C}(\nabla_X T)$ for any contraction $\mathcal{C}$.

Remark 2.2.13. Note that, for given $\psi \in \Gamma(E)$, we only need to know X_p to calculate the value of $\nabla_X\psi$ at some point $p \in M$. We use this to also write $\nabla_v\psi$ for a given $v \in T_pM$.

In the following, let ∇ be a connection over the vector bundle E with base M.

Definition 2.2.14. Let $I \subset \mathbb{R}$ be an open interval and $\gamma \colon I \to M$ a path. A *vector field along* γ is a map $\psi \colon I \to E$ such that $\pi \circ \psi = \gamma$. Thus, for any $s \in I$, $\psi(s)$ is an element of the vector space $E_{\gamma(s)}$. We write $\Gamma(\gamma^*E)$ for the $\mathfrak{F}(I)$-module of vector fields along γ.

Let $s \mapsto \gamma'(s) := \gamma_*(\partial_t|_s)$, $s \in I$, be the *tangent vector field* along γ. The *induced covariant derivative* $\nabla_{\gamma'} \colon \Gamma(\gamma^*E) \to \Gamma(\gamma^*E)$ is uniquely determined by the conditions:

1. $(\nabla_{\gamma'}(\Phi \circ \gamma))(s) = \nabla_{\gamma'(s)}\Phi$ for all $s \in I$ and $\Phi \in \Gamma(E)$,

2. $\nabla_{\gamma'}(f\psi) = \frac{\mathrm{d}f}{\mathrm{d}t}\psi + f\nabla_{\gamma'}\psi$ for all $f \in \mathfrak{F}(I)$ and $\psi \in \Gamma(\gamma^*E)$,

3. $\nabla_{\gamma'}(\psi + \phi)$ for all $\psi, \phi \in \Gamma(\gamma^*E)$.

A vector field ψ along γ is called *parallel* if $\nabla_{\gamma'}\psi = 0$ holds.

Let $s, t \in I$. The *parallel transport* along γ is the map $P(\gamma)_s^t \colon E_{\gamma(s)} \to E_{\gamma(t)}$, defined by $v \mapsto \psi(t)$, where ψ is the unique parallel vector field along γ with $\psi(s) = v$.

Definition 2.2.15. Let $X, Y \in \Gamma(TM)$ and $\psi \in \Gamma(E)$ be arbitrary vector fields. We define the *curvature operator* $\mathcal{R} = \mathcal{R}^{\nabla} \colon \Gamma(TM) \times \Gamma(TM) \times \Gamma(E) \to \Gamma(E)$ of ∇ by setting

$$\mathcal{R}(X, Y, \psi) =: \mathcal{R}(X, Y)\psi = \nabla_X(\nabla_Y\psi) - \nabla_Y(\nabla_X\psi) - \nabla_{[X,Y]}\psi.$$

Definition 2.2.16. We say that ∇ is *compatible* with a given vector bundle metric g if $\nabla_X g = 0$ for each $X \in \Gamma(TM)$, or equivalently, the product rule

$$X(g(\psi, \phi)) = g(\nabla_X\psi, \phi) + g(\psi, \nabla_X\phi)$$

holds for each $X \in \Gamma(TM)$, $\psi, \phi \in \Gamma(E)$.

2.2.3 Semi-Riemannian Manifolds and the Levi-Civita Connection.

In this subsection, we specialize to a connection over the tangent bundle TM that is uniquely determined by a metric structure:

Theorem 2.2.17. *Let (M, g) be a semi-Riemannian manifold. There exists a unique covariant derivation $\nabla = \nabla^g$ over TM that is compatible with g and has vanishing torsion, i.e., $\nabla_X Y - \nabla_Y X = [X, Y]$ holds for all $X, Y \in \Gamma(TM)$. This derivation is called the* Levi-Civita *connection of g.*

Remark 2.2.18. Given a one-form ℓ on (M, g), it may be shown that there exists a unique, torsion-free covariant derivative D such that $\mathrm{D}_X g = 2\ell(X)g$ holds for all vector fields X. This derivative is called the *Weyl connection* generated by ℓ [Fol70, Hal92, GNS11]. Obviously, the Levi-Civita connection is the unique Weyl connection generated by the zero one-form.

Remark 2.2.19. At this point, we would also like to remark on the topological obstructions for a manifold to admit a semi-Riemannian metric over its tangent bundle:

- Any manifold admits a Riemannian metric.

- A manifold M admits a Lorentzian metric if and only if it admits a nowhere vanishing vector field. This is the case if either M is non-compact, or M is compact and the Euler characteristic of M vanishes, see Remark 2.2.37 below.

- In general, a vector bundle E admits a metric of index ι if and only if it splits as $E \cong E_1 \oplus E_2$ where E_1 is a vector bundle of rank ι: if the bundle is trivial, metric tensors of arbitrary signature exist.

In the following, let (M, g) be a semi-Riemannian manifold of index ι and dimension d, and ∇ its Levi-Civita connection.

Definition 2.2.20. A path $\gamma \colon I \to M$ is called a *geodesic* if $\nabla_{\gamma'}\gamma' = 0$ holds. We say that a vector field is geodesic if its integral curves are geodesics.

Remark 2.2.21. In other words, the tangent vector field of a geodesic is preserved by parallel transport with respect to the Levi-Civita connection. The geodesic equation $\nabla_{\gamma'}\gamma' = 0$ is the Euler–Lagrange equation with respect to the action

$$S(\gamma) = \int_0^1 g_{\gamma(t)}(\gamma'(t), \gamma'(t))\, \mathrm{d}t. \tag{2.1}$$

Paths having a tangent vector field that is preserved by parallel transport with respect to a connection other than Levi-Civita are called *auto parallel*. Such paths are not stationary points of the action (2.1), in general.

Also note that we only discuss affinely parametrized geodesics. Any *pregeodesic* γ, satisfying $\nabla_{\gamma'}\gamma' = f\gamma'$ for some function f along γ, can be reparametrized to yield a geodesic as defined above.

Definition 2.2.22. A semi-Riemannian manifold is called *(geodesically) complete* if every geodesic $\gamma \colon I \to M$ may be extended to arbitrary parameter values, $I = \mathbb{R}$.

Remark 2.2.23. For example, any compact Riemannian manifold is complete. The Hopf–Rinow theorem [O'N83, p. 138] on the completeness of Riemannian manifolds has no direct analogue for general semi-Riemannian manifolds. Theorems on the timelike geodesic completeness of Lorentzian manifolds are known as singularity theorems.

Definition 2.2.24. For a function $f \in \mathfrak{F}(M)$, its *(exterior) derivative* or *differential* is given by the one-form $\mathrm{d}f$ with $\mathrm{d}f(X) = X(f)$ for $X \in \Gamma(TM)$. The *gradient* of f is defined as the vector field $\nabla f = (\mathrm{d}f)^\sharp$.

The *covariant differential* of a $(0, s)$-tensor field τ, $s \geq 1$, is given by the $(0, s + 1)$-tensor field $\nabla\tau$ with $(\nabla\tau)(X, X_1, \ldots, X_s) = (\nabla_X\tau)(X_1, \ldots, X_s)$ for $X, X_1, \ldots, X_s \in \Gamma(TM)$. If τ is skew-symmetric, i.e., an s-form, its *exterior derivative* is defined as the $s + 1$-form $\mathrm{d}\tau = s!\, \mathrm{asym}(\nabla\tau)$.

The *curl* of a vector field X is the two-form $\mathrm{curl}(X) := \mathrm{d}(X^\flat)$.

Definition 2.2.25. The *divergence* of a $(0, s)$-tensor field τ, $s \geq 1$, is given by the

$(0, s-1)$-tensor field $\delta\tau = \delta^g\tau$ with

$$(\delta\tau)(X_2, \ldots, X_s) = \text{Trace}(Z \mapsto ((\nabla_Z\tau)(\,\cdot\,, X_2, \ldots, X_s))^\sharp).$$

for vector fields $X_2, \ldots, X_s \in \Gamma(TM)$.

By definition, $\delta f = 0$ for any function f.

The divergence of a vector field X is the function $\text{div}(X) = \delta(X^\flat)$.

Definition 2.2.26. The *Hodge–Laplace operator* acting on s-forms is given by $\triangle = \text{d}\delta + \delta\text{d}$. If g is a Riemannian metric, we say that an s-form τ is *harmonic* if $\triangle\tau = 0$.

Remark 2.2.27. Obviously, $X(f) = \text{d}f(X) = g(\nabla f, X)$ are just different ways to write the derivative of a function f in the direction of a vector field X.

Note that $\text{d} \circ \text{d} = 0$. Also, for an s-form τ, $\delta = (-1)^{d(s+1)+\iota} * \text{d}*$, and consequently, $\delta \circ \delta = 0$ when acting on s-forms.

Remark 2.2.28. Writing $\nabla_k\tau_{i_1\ldots i_s} := (\nabla\tau)_{ki_1\ldots i_s}$, in local coordinates,

$$(\delta\tau)_{i_2\ldots i_s} = g^{kl}\nabla_k\tau_{li_2\ldots i_s} = \nabla_k\tau^k_{i_2\ldots i_s}, \quad \text{div}(X) = \nabla_i X^i,$$
$$(\text{d}\tau)_{ki_1\ldots i_s} = s!\nabla_{[k}\tau_{i_1i_2\ldots i_s]}, \quad (\text{curl}(X))_{ij} = \nabla_i X_j - \nabla_j X_i.$$

The exterior derivative may in fact be defined via ordinary coordinate derivatives:

$$(\text{d}\tau)_{ki_1\ldots i_s} = s!\partial_{[k}\tau_{i_1i_2\ldots i_s]}.$$

Thus, the exterior derivative on differential forms is a notion tied to the basic manifold structure. This can also be seen from the following index-free formula:

$$(\text{d}\tau)(X_1, \ldots, X_{s+1}) = \sum_{i=1}^{s+1}(-1)^{i+1}X_i(\tau(X_1, \ldots, \cancel{X_i}, \ldots, X_{s+1}))$$
$$+ \sum_{i<j}(-1)^{i+j}\tau([X_i, X_j], X_1, \ldots, \cancel{X_i}, \ldots, \cancel{X_j}, \ldots, X_{s+1}).$$

Proposition 2.2.29. *For a local coordinate frame* $(\partial_0, \ldots, \partial_{d-1})$ *we have*

$$\nabla_{\partial_i}\partial_j = \Gamma^k_{ij}\partial_k$$

with

$$\Gamma^k_{ij} = \frac{1}{2}g^{kl}(\partial_i g_{jl} + \partial_j g_{il} - \partial_l g_{ij}).$$

The functions Γ^k_{ij} *are called the* Christoffel symbols. *The covariant derivative of a tensor*

$$\tau = \tau^{j_1\ldots j_r}_{i_1\ldots i_s}\,\partial_{j_1} \otimes \cdots \otimes \partial_{j_r} \otimes \text{d}x^{i_1} \otimes \cdots \otimes \text{d}x^{i_s}$$

has the components:

$$\nabla_k T^{j_1\dots j_r}{}_{i_1\dots i_s} = \partial_k T^{j_1\dots j_r}{}_{i_1\dots i_s} + \Gamma^{j_1}_{km_1} T^{m_1\dots j_r}{}_{i_1\dots i_s} + \dots + \Gamma^{j_r}_{km_r} T^{j_1\dots m_r}{}_{i_1\dots i_s}$$
$$- \Gamma^{n_1}_{ki_1} T^{j_1\dots j_r}{}_{n_1\dots i_s} - \dots - \Gamma^{n_s}_{ki_s} T^{j_1\dots j_r}{}_{i_1\dots n_s}.$$

Definition 2.2.30. Let $V, W, X, Y \in \Gamma(TM)$ be arbitrary vector fields.

1. The *Riemann curvature tensor* $R = R^g \in \Gamma(TM^* \otimes TM^* \otimes TM^* \otimes TM^*)$ is defined as
$$R(V, W, X, Y) = g(\mathcal{R}^\nabla(V, W)X, Y).$$

2. The *Ricci curvature tensor* $\mathrm{Ric} = \mathrm{Ric}^g \in \Gamma(TM^* \otimes TM^*)$ is given by
$$\mathrm{Ric}(X, Y) = \mathrm{Trace}(Z \mapsto \mathcal{R}^\nabla(Z, X)Y).$$

3. The *scalar curvature* $\kappa = \kappa^g \in \mathfrak{F}(M)$ is defined as
$$\kappa = \mathrm{Trace}(Z \mapsto (\mathrm{Ric}(Z, \cdot))^\sharp).$$

4. Provided that $d \geq 3$, the *Weyl curvature tensor* $C = C^g \in \Gamma(TM^* \otimes TM^* \otimes TM^* \otimes TM^*)$ is defined as
$$C = R - \frac{1}{d-2}\left(\mathrm{Ric} - \frac{\kappa}{d}g\right) \otimes g - \frac{\kappa}{2d(d-1)}g \otimes g,$$

where "$\otimes$" denotes the *Kulkarni–Nomizu product* of two symmetric $(0, 2)$-tensors A and B:
$$(A \otimes B)(V, W, X, Y) := A(V, X)B(W, Y) - A(V, Y)B(W, X)$$
$$+ A(W, Y)B(V, X) - A(W, X)B(V, Y).$$

We set $C = 0$ for $d = 0, 1, 2$.

Remark 2.2.31. We have $R = 0$ for $d = 0, 1$ and $R = \frac{\kappa}{2d(d-1)}g \otimes g$ for $d = 2$. For $d = 3$, one always has $C = 0$.

Remark 2.2.32. When we work in local coordinates, we adopt the usual Ricci calculus notation and denote R, Ric and κ by the same letter R. These objects are then identified by the number of indices: $(\mathrm{Ric})_{ij} = R_{ij} = R^m{}_{imj}$, $\kappa = R = R^m{}_m$.

Definition 2.2.33. We say that (M, g) is

1. a space of *constant curvature* if $R = k_1 g \otimes g$ (*flat* if $k_1 = 0$) for some constant $k_1 \in \mathbb{R}$,

2. an *Einstein space* if $\mathrm{Ric} = k_2 g$ (*Ricci-flat* if $k_2 = 0$) for some constant $k_2 \in \mathbb{R}$,

3. a space of *constant scalar curvature* if κ is constant.

Remark 2.2.34. For manifolds with special curvature structure we have the following:

- Every space of constant curvature is Einstein, and every Einstein space has constant scalar curvature. One necessarily has $k_1 = (2d(d-1))^{-1}\kappa$ and $k_2 = d^{-1}\kappa$.

- If $d = 2$, the three notions are equivalent. In $d = 3$ dimensions, a manifold is an Einstein space if and only if it is of constant curvature ([Bes87, Proposition 1.120]). In the literature, one often assumes $d \geq 3$ or even $d \geq 4$ when investigating Einstein manifolds.

- For connected manifolds of dimension $d \geq 3$, one may drop the assumption that the factor k_1 or k_2 is a constant; this fact is then directly implied by the *contracted Bianchi identity* $\delta(\mathrm{Ric}) = \frac{1}{2}\mathrm{d}\kappa$.

- In the context of relativity theory, Einstein Lorentzian manifolds are solutions to the vacuum Einstein–Hilbert equation with cosmological constant Λ, $\mathrm{Ric} - \frac{\kappa}{2}g + \Lambda g = 0$.

Theorem 2.2.35. *If (M, g) is a simply connected and complete Riemannian manifold of constant curvature, it is isometric to a space form: a round sphere, a hyperbolic space, or Euclidean space.*

Proof. See [Lee97, Theorem 11.2]. Note that we adopt the convention that simply connected spaces are, by definition, path connected. □

2.2.4 Differential Forms. Integration.

In the following, let M be a manifold of dimension d, and denote the vector space of k-forms, i.e., skew-symmetric covariant tensors of rank k on M, by $\Lambda^k(M)$. We understand $\Lambda^{-1}(M) \cong \mathbb{R}$ as the space of constant functions on M. Useful texts on the following topics and beyond are, for example, [YB53, HR72, Aub82, dR84, HO03].

Definition 2.2.36. Let $k \in \{0, \ldots, d\}$. A k-form τ is *closed* if $\mathrm{d}\tau = 0$. It is *exact* if there exists a $k-1$-form η with $\tau = \mathrm{d}\eta$. Note that every exact form is closed. Define the quotient vector space

$$H^k(M) = \frac{\{\tau \in \Lambda^k(M)|\tau \text{ is closed}\}}{\{\tau \in \Lambda^k(M)|\tau \text{ is exact}\}}.$$

This is the k-th *de Rham cohomology group*, $B_k(M) = \dim(H^k(M))$ is the k-th *Betti number* of M (which is either a natural number or equal to ∞).

Remark 2.2.37. We gather some useful facts:

- Obviously, every closed k-form on M is exact if and only if $B_k(M) = 0$ holds.

- We have that $B_0(M)$ is the number of connected components of M.

- If M is contractible, $B_k(M) = 0$ holds for every $k = 1, \ldots, d$. This is Poincaré's lemma (see below).

- If M is simply connected, M is orientable [Nar85, Corollary 2.7.6] and $B_1(M) = 0$ holds [Lee02, Theorem 15.17]. The converse is not true – the Poincaré spaces [ST80, Section 62 ff.] provide infinitely many counter-examples. Nevertheless, simply connected manifolds may be thought of as the standard examples of orientable manifolds with vanishing first Betti number.

- A manifold is called *of finite type* if it admits a finite cover $\{U_1, \ldots, U_N\}$ such that all non-empty intersections $\bigcap_j U_j$ are diffeomorphic to $\mathbb{R}^d$. Any compact manifold is of finite type. If M is of finite type, $B_k(M) < \infty$ holds for every $k = 0, \ldots, d$ [BT82].

- For a product manifold $M = N_1 \times N$, where N_1 is of finite type,

$$H^k(M) \cong \bigoplus_{p+q=k} H^p(N_1) \otimes H^q(N).$$

This is the *Künneth formula*. In particular, if $I \subset \mathbb{R}$ is an open interval, $H^k(I \times N) \cong H^k(N)$.

- **Poincaré duality.** If M is compact and orientable, $B_k(M) = B_{d-k}(M)$.

- If M is of finite type, the *Euler characteristic* of M is given by

$$\chi(M) = \sum_{k=0}^{d} (-1)^k B_k(M).$$

If M is compact, this number vanishes if and only if there exists a vector field on M without zeros. (Any non-compact manifold admits such a vector field.) If M is compact, orientable and of odd dimension, via Poincaré duality we have that $\chi(M) = 0$.

Furnish M with a Riemannian metric g.

- If M is compact and orientable, $H^k(M)$ is isomorphic to the space of harmonic k-forms on M.

- If M is compact and orientable, and (M, g) has positive Ricci curvature, then the first Betti number of M vanishes [Boc46]. More generally, if for each $p \in M$ the quadratic form $F\colon \bigwedge^k T_p M \to \mathbb{R}$ defined by

$$F(\tau) = R_{ij}\tau^{ii_2\ldots i_k}\tau^j_{\ i_2\ldots i_k} + \tfrac{k-1}{2}R_{ijmn}\tau^{iji_3\ldots i_k}\tau^{mn}_{\quad i_3\ldots i_k},$$

is positive definite, then $B_k(M) = 0$ [YB53, Theorem 3.4].

Theorem 2.2.38. Poincaré lemma. *Any k-form on M, k ≥ 1, is closed if and only if it is locally exact. The result holds globally if M is contractible.*

In the following, let (M, g) be a semi-Riemannian manifold of dimension d. For a function $f \in \mathfrak{F}(M)$ with compact support contained in a chart (U, ϕ), the integral with respect to the metric volume element is given by

$$\int_M f \, \mathrm{dvol}(g) = \int_{\phi(U)} f \circ \phi^{-1} \sqrt{|\det(g_{ij})|} \, \mathrm{d}x_1 \cdots \mathrm{d}x_d.$$

For an arbitrary function with compact support, the integral is defined by summation with respect to a partition of unity. If M is compact, its volume is given by $\mathrm{vol}(M) = \mathrm{vol}^g(M) = \int_M \mathrm{dvol}(g)$.

Definition 2.2.39. The L^2-*inner product* is given by

$$\langle\!\langle \tau_1, \tau_2 \rangle\!\rangle = \int_M \langle \tau_1, \tau_2 \rangle \, \mathrm{dvol}(g),$$

where τ_1 and τ_2 are tensor fields on M of the same rank, and τ_1 or τ_2 has compact support.

Theorem 2.2.40. Green's integral identity. *Let η be a $k-1$-form and τ a k-form on M. Suppose M is orientable, and η or τ has compact support. Then,*

$$\langle\!\langle \mathrm{d}\eta, \tau \rangle\!\rangle + \langle\!\langle \eta, \delta\tau \rangle\!\rangle = 0.$$

Remark 2.2.41. In other words, thought of as acting on the direct sum $\bigoplus_k \Lambda^k(M)$, the differential operator $-\delta$ is the formal adjoint of d with respect to $\langle\!\langle \cdot, \cdot \rangle\!\rangle$. Note that the above formula makes sense since $\mathrm{d}\tau$ (or $\delta\eta$) has compact support whenever τ (or η) has compact support.

If we set $\eta = 1$, we recover a particular case of *Gauss' theorem*:

$$0 = \langle\!\langle \eta, \delta\tau \rangle\!\rangle = \int_M \delta\tau \, \mathrm{dvol}(g) = \int_M \mathrm{div}(X) \, \mathrm{dvol}(g),$$

where $X = \tau^\sharp$.

Now assume that g is a Riemannian metric.

Definition 2.2.42. Let $q \in \mathbb{R}$, $q \geq 1$. For any tensor field τ on M with compact support, define its L^q-*norm* by

$$\|\tau\|_{L^q} = \|\tau\|_{L^q(M,g)} = \left(\int_M |\tau|^q \, \mathrm{dvol}(g) \right)^{\frac{1}{q}}.$$

Furthermore, set

$$\|\tau\|_{L^\infty} = \|\tau\|_{L^\infty(M,g)} = \sup_{p \in M} |\tau_p|.$$

Remark 2.2.43. Obviously, $\|\tau\|_{L^2} = \sqrt{\langle\!\langle \tau, \tau \rangle\!\rangle}$.

More generally, locally integrable functions (for $q = \infty$, functions which are bounded almost everywhere) that are not necessarily smooth or with compact support but with finite L^q-norm may be used to construct the function spaces $L^q(M)$ in the usual way. The Sobolev space $W^{1,q}(M)$ is the completion of the space of differentiable L^q-functions with L^q-derivative, with respect to the norm $\|f\|_{W^{1,q}} = \|f\|_{L^q} + \|\mathrm{d}f\|_{L^q}$. We denote the corresponding spaces of k-forms with coefficients in $L^q(M)$ ($W^{1,q}(M)$) by $L^q_k(M)$ ($W^{1,q}_k(M)$). For more details on the subject, see [Aub82, GT06].

2.3 Conformal and Causal Geometry

Conformal transformations of Riemannian manifolds may be understood as generalizations of isometries: While isometries preserve the length of tangent vectors and angles between them, conformal transformations are only required to preserve angles. The classic example of conformal maps are holomorphic functions $f \colon \mathbb{C} \to \mathbb{C}$ with non-vanishing derivative f': for each point $p \in \mathbb{C}$, $f'(p)$ acts as a rotation-dilation $z \mapsto f'(p) \cdot z$.

In comparison, there is no general notion of an "angle" between tangent vectors of semi-Riemannian manifolds. However, conformal transformations may be analogously defined as a pointwise rescaling of the metric tensor which leaves invariant the causal character of tangent vectors, curves and submanifolds.

2.3.1 Conformal Transformations.

Definition 2.3.1. For a semi-Riemannian manifold (M, g) and a function $\phi \in \mathfrak{F}(M)$, we define a *conformal change* of the metric g as the map $g \mapsto \tilde{g} := \mathrm{e}^{2\phi} g$. We say that two metrics g and $\tilde{g}$ are *conformally related* if there exists a conformal change mapping g to $\tilde{g}$.

We say that the semi-Riemannian manifolds (M, g) and (M_1, g_1) are *conformally equivalent* if there exists a diffeomorphism $f \colon M \to M_1$ such that g and f^*g_1 are conformally related. The map f is called a *conformal transformation*. It is called an *isometry* if $g = f^*g_1$.

A map $f \colon M \to M_1$ is called a *local* conformal transformation (local isometry) if for each point $p \in M$ there exists a neighborhood U of p and a neighborhood V of $f(p)$ such that $f_{|U} \colon U \to V$ is a conformal transformation (an isometry).

Remark 2.3.2. We briefly discuss manifolds that are *conformally flat*, i.e., conformally equivalent to a flat manifold. Let (M, g) be a Riemannian or Lorentzian manifold with $\dim(M) = d$. If $d = 2$, then (M, g) is always locally conformally flat: about each point there exist coordinates (x, y), called *isothermal coordinates*, such that the metric takes the form $g = \mathrm{e}^{2\phi}(\pm \mathrm{d}x^2 + \mathrm{d}y^2)$ for some function ϕ. The global view depends on the signature of the metric:

- In the Riemannian case, the Poincaré uniformization theorem ensures that every compact surface may be conformally mapped onto a surface of constant curvature. Furthermore, every non-compact simply connected Riemannian surface is conformally equivalent to the hyperbolic or Euclidean plane.

- Compact Lorentzian manifolds must have vanishing Euler characteristic which implies that any compact connected Lorentzian surface is diffeomorphic to the torus or the Klein bottle. Since the Gauss–Bonnet theorem also holds for indefinite metrics [BN84], the only Lorentzian surfaces of constant curvature are therefore the flat ones. However, there are many Lorentzian surfaces which are not conformally flat [Sán97]. As for the non-compact case, instead of two as in the Riemannian case, there exist infinitely many conformal classes. Moreover, the family of conformal classes depends on the differentiability of the conformal mappings in question [Wei96].

If $d = 3$, then (M, g) is locally conformally flat if and only if the *Cotton tensor* Υ vanishes, which is given by

$$\Upsilon(X, Y, Z) = (\nabla_Z \mathrm{Ric})(X, Y) - (\nabla_Y \mathrm{Ric})(X, Z)$$
$$+ \frac{1}{4}(\mathrm{d}\kappa(Y)g(X, Z) - \mathrm{d}\kappa(Z)g(X, Y))$$

for $X, Y, Z \in \Gamma(TM)$. If $d \geq 4$, then (M, g) is locally conformally flat if and only if the Weyl curvature tensor vanishes.

The following definition is not part of the standard literature but suggests itself:

Definition 2.3.3. Let M be a manifold, $\phi \in \mathfrak{F}(M)$, and P a vector bundle metric over some sub-bundle F of TM, i.e., $TM = E \oplus F$ for some other bundle E. We define a *conformal change* of the vector bundle metric P as the map $P \mapsto \tilde{P} := \mathrm{e}^{2\phi}P$. We say that two vector bundle metrics P and $\tilde{P}$ over F are *conformally related* if there exists a conformal change mapping P to $\tilde{P}$.

A *conformal transformation* of the pair (M, P) is a diffeomorphism $f \colon M \to M$ such that f preserves fibers, i.e., $f_{*p}F_p = F_{f(p)}$ for all $p \in M$, and $f^*P = \mathrm{e}^{2\phi}P$ for some function $\phi \in \mathfrak{F}(M)$. If $\phi = 0$, we say that f is an *isometry* of (M, P).

In the following, let (M, g) be a semi-Riemannian manifold with $\dim(M) = d$, and $\phi \in \mathfrak{F}(M)$.

Proposition 2.3.4. *Under the conformal change* $\tilde{g} = e^{2\phi}g$,

$$\tilde{\nabla}_Y X = \nabla_Y X + d\phi(Y)X + d\phi(X)Y - g(X,Y)\nabla\phi,$$
$$\tilde{\nabla}\xi = \nabla\xi - d\phi \vee \xi + \xi(\nabla\phi)g, \quad \widetilde{\operatorname{div}}X = \operatorname{div}X + d \cdot d\phi(X)$$

for any $X, Y \in \Gamma(TM)$ *and* $\xi \in \Gamma(TM^*)$. *The Levi-Civita derivation and divergence with respect to* $\tilde{g}$ *are denoted by* $\tilde{\nabla}$ *and* $\widetilde{\operatorname{div}}$, *respectively.*

Proof. The first formula can be found in [Bes87, Theorem 1.159], for example. In a local coordinate basis, it is equivalent to the following transformation formula for the Christoffel symbols:

$$\tilde{\Gamma}^k{}_{ij} = \Gamma^k{}_{ij} + \delta^k{}_i \partial_j \phi + \delta^k{}_j \partial_i \phi - g_{ij}\nabla^k\phi,$$

where $\delta^k{}_i = g^{kl}g_{li}$ is the Kronecker symbol. The other expressions can be easily computed from this formula:

$$\tilde{\nabla}_i \xi_j = \nabla_i \xi_j - 2\xi_{(i}\partial_{j)}\phi + g_{ij}\xi_k \nabla^k\phi, \quad \tilde{\nabla}_k X^k = \nabla_k X^k + d \cdot X^k \partial_k \phi.$$

$\square$

Remark 2.3.5. Let D be a Weyl connection generated by a one-form ℓ. Given a conformal change $\tilde{g} = e^{2\phi}g$, for all $X \in \Gamma(TM)$ it holds that $D_X\tilde{g} = 2\tilde{\ell}(X)\tilde{g}$ where $\tilde{\ell} = \ell + d\phi$.

The following formulas are also well-established; see [Bes87, Theorem 1.159]:

Proposition 2.3.6. *Under the conformal change* $\tilde{g} = e^{2\phi}g$, *the Riemann curvature tensor* R, *Weyl curvature tensor* C, *Ricci curvature tensor* Ric, *and the scalar curvature* κ *of* (M, g) *transform as follows:*

$$e^{-2\phi}\tilde{R} = R - g \otimes (\nabla d\phi - d\phi \otimes d\phi + \tfrac{1}{2}g(\nabla\phi, \nabla\phi)g), \quad e^{-2\phi}\tilde{C} = C,$$
$$\tilde{\operatorname{Ric}} = \operatorname{Ric} - (d-2)(\nabla d\phi - d\phi \otimes d\phi) - (\triangle\phi + (d-2)g(\nabla\phi, \nabla\phi))g,$$
$$e^{2\phi}\tilde{\kappa} = \kappa - 2(d-1)\triangle\phi - (d-2)(d-1)g(\nabla\phi, \nabla\phi).$$

Finally, semi-Riemannian manifolds may "look the same about every point":

Definition 2.3.7. We say that (M, g) is *(locally) homogeneous* if for every pair of points $p, q \in M$ there exists an isometry (a local isometry) $\phi\colon M \to M$ with $\phi(p) = q$.

Remark 2.3.8. In general, a locally homogeneous manifold need not be locally isometric to any homogeneous manifold. However, any locally homogeneous Riemannian manifold which is complete, or has a dimension of at most four, is locally isometric to a homogeneous manifold [Tri92, Pat96].

2.3.2 Causal Structure.

Since a semi-Riemannian g is not necessarily definite, we may distinguish tangent vectors by their *causal character*:

Definition 2.3.9. Let (M, g) be a semi-Riemannian manifold and $p \in M$. A tangent vector $v \in T_p M$ is called

1. *timelike* if $g_p(v, v) < 0$,

2. *null* if $g_p(v, v) = 0$ and $v \neq 0$,

3. *spacelike* if $g_p(v, v) > 0$ or $v = 0$,

4. *cospacelike* if for all $w \in T_p M$, $g_p(v, w) = 0$ implies that w is spacelike.

A vector field $X \in \Gamma(TM)$ is called timelike (null, spacelike, cospacelike) if $X_p \in T_p M$ is timelike (null, spacelike, cospacelike) for each $p \in M$.

Cospacelike vectors only appear for metrics of Riemannian or Lorentzian signature. In the following, let (M, g) be a Lorentzian manifold. Null tangent vectors are also called *lightlike*, and non-spacelike vectors are also called *causal* in this case.

Definition 2.3.10. Given a timelike vector field X on (M, g) we say that (M, g) is *time-oriented* by X: A non-spacelike tangent vector $v \in T_p M$ is called *future-pointing* (*past-pointing*) if $g_p(X_p, v) < 0$ $(g_p(X_p, v) > 0)$. Granted that a timelike vector field exists, (M, g) is *time-orientable*.

We say that a non-spacelike vector field Z is *future-pointing* (*past-pointing*) if Z_p is *future-pointing* (*past-pointing*) for every $p \in M$.

We call any time-oriented Lorentzian manifold of arbitrary dimension $d \geq 2$ a *spacetime*.

Remark 2.3.11. A time-orientation is a choice of lightcone in each tangent space that varies smoothly across the manifold. Any Lorentzian manifold admits a twofold time-orientable Lorentz covering. Any simply connected Lorentzian manifold is time-orientable [O'N83, p. 194]. In the following, we assume that (M, g) is time-oriented.

Definition 2.3.12. Let $I \subset \mathbb{R}$ be an open interval. A path $\gamma: I \to M$ is called timelike (null, spacelike) if the tangent vector $\gamma'(s)$ is timelike (null, spacelike) for each $s \in I$. A non-spacelike path is said to be *future-directed* (*past-directed*) if its tangent vector is future-pointing (past-pointing) everywhere.

Remark 2.3.13. Observe that the causal character of a tangent vector v is invariant against rescalings by a non-zero factor. It is thus a property of the line spanned by v in tangent space. In particular, the causal character of a path γ is preserved by reparametrizations and is thus a property of the curve traced by γ, i.e., its image.

The property to be future- or past-directed is invariant with respect to orientation-preserving reparametrizations. In most contexts, we may use the terms "path" and "curve" interchangeably.

Also note that with our conventions, non-spacelike curves are always *regular*, i.e., the tangent vector vanishes nowhere along the curve.

The following definition is consistent with Definition 2.3.9.

Definition 2.3.14. Let $p \in M$. A linear subspace $\Xi \subset T_p M$ is called timelike (null, spacelike) if the inner product $g_p(\,\cdot\,,\,\cdot\,)$ restricted to Ξ is non-degenerate of index 1 (degenerate, positive definite).

A submanifold $i \colon S \to M$ is timelike (null, spacelike) if $T_q S \cong i_{q*}(T_q S) \subset T_q M$ is timelike (null, spacelike) for all $q \in S$.

Definition 2.3.15. A Lorentzian metric g' on M is said to be *strictly wider* than g if its lightcones are strictly wider, i.e., for any $p \in M$ and tangent vector $v \in T_p M$, $g_p(v, v) \leq 0$ implies $g'_p(v, v) < 0$. We say that (M, g) is

1. *chronological* if no closed curve in M is timelike,

2. *causal* if no closed curve in M is causal,

3. *stably causal* if there exists a Lorentzian metric g' strictly wider than g such that (M, g') is causal.

Remark 2.3.16. Observe that

$$(M, g) \text{ is stably causal } \Rightarrow (M, g) \text{ is causal } \Rightarrow (M, g) \text{ is chronological.}$$

Spacetimes which contain a closed timelike curve through every point are also called *totally vicious*.

Proposition 2.3.17. *If M is compact, (M, g) is non-chronological.*

Proof. See [BEE96, Proposition 3.10]. □

Remark 2.3.18. Although this might be considered an intuitive conjecture, compact connected spacetimes need not be totally vicious [Mat87].

Let $\mathrm{Lor}(M)$ be the set of Lorentzian metrics on M. Fix an atlas that is countable and locally finite. For every continuous function $\delta \colon M \to \mathbb{R}$, $\delta > 0$, and $g \in \mathrm{Lor}(M)$ define

$$U_\delta(g) = \{\, g' \in \mathrm{Lor}(M) \,|\, |g - g'| < \delta \,\},$$

where $|g - g'| < \delta$ means that at each point $p \in M$, we have $\max_{h,k} \left| (g_p - g'_p)_{hk} \right| < \delta(p)$ with respect to every chart that contains p. The *fine C^0-topology* on $\mathrm{Lor}(M)$ is the topology generated by the sets of the form $U_\delta(g)$. This topology does not depend on the atlas used.

Theorem 2.3.19. *The following statements are equivalent:*

1. (M, g) *is stably causal.*

2. *There exists a continuous function on (M, g) which increases along every future-directed timelike curve, i.e., there exists a* time function.

3. *There exists a smooth function on (M, g) with past-pointing timelike gradient, i.e., there exists a* temporal function.

4. *There exists a fine C^0-neighborhood of g which only contains metrics g' such that (M, g') causal.*

Proof. See [BS05] and the references therein. The last characterization can be found in [BEE96]. Note that any temporal function is also a time function, but even smooth time functions are not necessarily temporal. $\square$

Remark 2.3.20. Especially in the physics literature [Mai66, Obu00], there has been the following argument to prove the absence of closed causal curves completely contained in a coordinate neighborhood U: Suppose we have coordinates $(t, x_1, \ldots, x_{d-1})$ on U. The metric coefficients g_{ij} with respect to these coordinates may be written in the following partitioned matrix form:

$$(g_{ij}) = \begin{pmatrix} u_0 & \vec{u}^T \\ \vec{u} & P \end{pmatrix},$$

where P is a square matrix with $d - 1$ columns/rows.

Now, assume that there is a closed causal curve $\gamma \colon I \to M$ such that $\gamma(I)$ is contained in U. Since γ is closed, there exist $a, b \in I$ with $\gamma(a) = \gamma(b)$, therefore $(t \circ \gamma)(a) = (t \circ \gamma)(b)$ holds. Thus, there exists some $s \in [a, b]$ with $(t \circ \gamma)'(s) = 0$, i.e., the first component of the tangent vector vanishes, $(\gamma'(s))^i = (0, \vec{\gamma}'(s))$. Since γ is causal, we have $(p = \gamma(s))$:

$$0 \geq g_p(\gamma'(s), \gamma'(s)) = g_{ij}|_p (\gamma'(s))^i (\gamma'(s))^j = (\vec{\gamma}')^T(s) P_p \vec{\gamma}'(s).$$

Thus, if P is positive definite everywhere on U, no closed causal curve may be contained in U. Remember that P is positive definite if and only if all leading principal minors of P are positive (Sylvester's criterion). In fact, we have that $\det(P) > 0$ already implies that t is a temporal function on U, strengthening the above criterion considerably: First, via formulas for the inverse and determinant of partitioned matrices (see [HS81]) we have that $\det(g) = u_0 \det(P) - \vec{u}^T \mathrm{adj}(P)\vec{u}$ (assuming $d \geq 3$). Since g is a Lorentzian metric, $\det(g) < 0$, thus $u_0 < \vec{u}^T P^{-1} \vec{u}$ holds. Therefore,

$$g(\nabla t, \nabla t) = g^{-1}(\mathrm{d}t, \mathrm{d}t) = (u_0 - \vec{u}^T P^{-1} u)^{-1} < 0.$$

Conversely, if f is a temporal function on M, about each point there exists a coordinate neighborhood U and coordinates (t, x) such that $t = f|_U$, see [Bee76, Lemma 7]. As for the case $d = 2$, any spacetime homeomorphic to $\mathbb{R}^2$ is stably causal [BEE96, Theorem 3.43].

Proposition 2.3.21. *Let (M_1, g_1) be a Lorentzian manifold conformally equivalent to (M, g). Then, (M_1, g_1) is chronological (causal, stably causal) if and only if (M, g) is chronological (causal, stably causal).*

2.4 Relativity Theory and Cosmology

Introductory texts on relativity theory and cosmology are, for example, [HE73, SW77, Ray79, Wal84, Jos93, Thi97, Car03, Isl04, Ste04, PK06, CB09]. The general theory of relativity is a physical theory which describes gravitation, one of the fundamental interactions between matter and energy, at macroscopic scales. It states that the gravitational field is in fact the curvature of a geometry which represents a synthesis of space and time, called *spacetime*. More precisely, spacetime is modeled by a Lorentzian manifold (M, g), and matter/energy couples via the Ricci part of the curvature, where this coupling is quantified by the Einstein–Hilbert equation

$$\mathrm{Ric} - \frac{\kappa}{2} g = k_d T,$$

where $T \in \Gamma(TM^* \vee TM^*)$ is the *energy–momentum tensor* and k_d is a coupling constant, depending on the spacetime dimension d [CF91]. Throughout this work we assume that the coupling constant is absorbed in T and set $k_d = 1$. (Two-dimensional spacetimes are always "empty", i.e., $T = 0$.) The Weyl part of the curvature may be interpreted as purely gravitational degrees of freedom. In conventional relativity theory, the spacetime is usually assumed to be of dimension $d = 4$ – however, with the advent of theories presuming extra spatial dimensions like M-theory and supergravity [BBS06] or Randall–Sundrum models [RS99a, RS99b], Einstein gravity and Lorentzian geometry for $d \neq 4$ attracted an increasing interest also from physicists. It is often stated that the left-hand side of the equation represents the spacetime curvature, whereas the right-hand side represents the matter content of this spacetime. It should be noted, however, that the energy–momentum tensor often depends on the metric as geometric data as well, and one should thus be careful with such an interpretation. Also note that the theory gives surprisingly little a priori information about how the topology of M might look like.

Furthermore, the motion of free-falling test particles through spacetime is governed by the geodesic equations. More precisely, free-falling particles with non-vanishing mass travel along timelike geodesics, whereas massless particles, such as photons, move along null geodesics. In general, a timelike path represents an observer's *world line* through spacetime. The outcomes of measurements of physical quantities perceived by a specific observer in a Newtonian sense are most conveniently calculated by projecting a relativistic quantity onto the observer's *restspace* which is the orthogonal complement of the world line's tangent space. For $d = 4$, this procedure is called $(3 + 1)$-formalism or $(3 + 1)$-decomposition, and is described in more detail in the next chapter.

Since gravitation is responsible for the formation of large scale structures in the universe, relativity theory may not only be applied to describe small isolated gravitational systems such as stars or black holes but also to describe these large scale structures, and in fact to describe the largest possible structure, the universe itself. Also, relativity theory treats space and time in a unified manner, thus a model of the universe necessarily is also a model of its history.

From the theoretical physics point of view, this work is concerned with exact solutions to the Einstein–Hilbert equation [KSMH79]. Since one may take any Lorentzian manifold (M, g) and claim that $\mathrm{Ric} - \frac{\kappa}{2}g$ is the energy–momentum tensor, there are conceptual difficulties as to what constitutes a physically meaningful exact solution. Several conditions may be applied: energy conditions, equations of state, asymptotic flatness for solutions describing isolated stellar bodies, certain symmetries for cosmological models (Copernican principle). Here, we adopt the simplified view that an exact solution is a spacetime where the energy–momentum tensor is of a particular form. We formally define such models later in Chapters 5 and 6: the viscous fluid and the charged perfect fluid.

Chapter 3

KINEMATICAL MANIFOLDS

A frequently used technique in relativity theory is the splitting of certain tensors into components parallel or perpendicular to a given timelike vector field, which is called reference frame or observer field in this context. This vector field is usually distinguished from other vector fields by some convenient property. For example, it might be proportional to a Killing vector field or an eigenvector field of the energy–momentum tensor.

Definition 3.0.1. Let (M, g) be a Riemannian or Lorentzian manifold. A cospacelike unit vector field V ($\varepsilon := g(V, V) = \pm 1$) on (M, g) is called a *reference frame* or *observer field* on (M, g). The triple (M, g, V) is a *kinematical manifold*, or, in the Lorentzian case, a *kinematical spacetime*, *world model* or *cosmological model*. An integral curve of V is called a *standard observer*.

In the following sections, for the remainder of this chapter, let V be a reference frame on a Riemannian or Lorentzian manifold (M, g) of dimension $d = D + 1$. Later on, our main focus is the Lorentzian case, where we assume (M, g) to be time-oriented by V. The nowhere vanishing vector field V induces an orthogonal tangent bundle splitting $TM = VM \oplus HM$ since for any $p \in M$, any tangent vector $v \in T_p M$ may be uniquely decomposed as the sum of a *vertical vector* $v^{\parallel} \in V_p M$ parallel to V_p and a *horizontal vector* $v^{\perp} \in H_p M$ that is g_p-orthogonal to V_p.

The tensor field $P = g - \varepsilon V^{\flat} \otimes V^{\flat}$ is a Riemannian metric over HM (the *transversal metric*), and $v \mapsto P_p(v, \cdot)^{\sharp} = v^{\perp}$ is the projection onto the restspace. For a function $\phi \in \mathfrak{F}(M)$, we define a conformal change of a kinematical manifold as the transformation that maps (M, g, V) to $(M, \tilde{g}, \tilde{V})$, where $\tilde{g} = \mathrm{e}^{2\phi} g$ and $\tilde{V} = \mathrm{e}^{-\phi} V$. This mapping induces a conformal change of the transversal metric: $P \mapsto \tilde{P} = \mathrm{e}^{2\phi} P$.

3.1   Reference Frames and $(D + 1)$-Decomposition

For an overview on the application of $(3 + 1)$-decomposition to cosmology we refer to [Ev99].

Proposition 3.1.1. *For any $(0,2)$-tensor field τ on M there exists the unique $(D+1)$-decomposition*

$$\tau = \alpha V^\flat \otimes V^\flat + r \otimes V^\flat + V^\flat \otimes s + \Sigma + \Omega + \frac{\beta}{d-1}P, \tag{3.1}$$

where $\alpha, \beta \in \mathfrak{F}(M)$, $r, s \in \Gamma(HM^*)$, $\Sigma \in \Gamma(HM^* \vee HM^*)$ is trace-free and $\Omega \in \Gamma(HM^* \wedge HM^*)$. These components may be determined via the formulas

$$\alpha = \tau(V,V), \quad \beta = \mathrm{tr}(\tau) - \varepsilon\alpha,$$
$$r = \varepsilon\tau(\,\cdot\,,V) - \alpha V^\flat, \quad s = \varepsilon\tau(V,\,\cdot\,) - \alpha V^\flat,$$
$$\Sigma = \mathrm{sym}(\tau) - \frac{1}{2}(r+s) \vee V^\flat - \frac{\beta}{d-1}P - \alpha V^\flat \otimes V^\flat,$$
$$\Omega = \mathrm{asym}(\tau) - \frac{1}{2}(r-s) \wedge V^\flat.$$

Proof. It is not difficult to check that the defined quantities have the desired properties and yield the decomposition. As for uniqueness, suppose there are α', β', r', s', Σ', Ω' that also satisfy the stated requirements. Then,

$$0 = (\alpha - \alpha')V^\flat \otimes V^\flat + (r - r') \otimes V^\flat + V^\flat \otimes (s - s')$$
$$+ (\Sigma - \Sigma') + (\Omega - \Omega') + \frac{\beta - \beta'}{d-1}P.$$

Inserting V into both slots on the right-hand side yields $\alpha = \alpha'$; inserting V and arbitrary $X \in \Gamma(HM)$ gives $r = r'$, $s = s'$. Taking the trace implies $\beta = \beta'$. Finally, taking the symmetric and skew-symmetric part yields $\Sigma = \Sigma'$ and $\Omega = \Omega'$, respectively. $\qquad\square$

Remark 3.1.2. In terms of a local frame, this decomposition may also be easily understood as follows: Complete $V^\flat$ to a dual orthonormal frame $(V^\flat, E^1, \ldots, E^{d-1})$. Any $(0,2)$-tensor τ may then be written as

$$\tau = \alpha V^\flat \otimes V^\flat + r_i E^i \otimes V^\flat + s_i V^\flat \otimes E^i + (\tau^\perp)_{ij} E^i \otimes E^j$$

where summation over $i,j = 1, \ldots, d-1$ is understood. If we define $r = r_i E^i$, $s = s_i E^i$, Σ as the symmetric, trace-free part and Ω as the skew-symmetric part of $\tau^\perp = (\tau^\perp)_{ij} E^i \otimes E^j$, this yields the above decomposition.

Example 3.1.3. If F is a skew-symmetric $(0,2)$-tensor field, the $(D+1)$-decomposition of F, $D \geq 2$, may be written in the form

$$F = E \wedge V^\flat + *B,$$

where $E \in \Gamma(HM^*)$, $B \in \Gamma(\bigwedge^{d-3} HM^*)$ and "$*$" denotes the Hodge dual over HM with respect to P. Note that in the case $d = 4$, B is a horizontal one-form. In the Lorentzian case, if F may be physically interpreted as an electromagnetic field tensor, E represents the *electric field* and B the *magnetic field* measured by a standard observer.

Example 3.1.4. If T is a symmetric $(0,2)$-tensor field, the $(D+1)$-decomposition of T may be written in the form

$$T = \rho V^\flat \otimes V^\flat + p^* P + q \vee V^\flat + \Pi,$$

where $\rho, p \in \mathfrak{F}(M)$, $q \in \Gamma(HM^*)$, and $\Pi \in \Gamma(HM^* \vee HM^*)$ is trace-free. In the Lorentzian case $\varepsilon = -1$, if T represents the energy–momentum tensor of a fluid, we have the following terminology or physical interpretation of the components:

$$\rho = T(V,V), \qquad\qquad\qquad\qquad\qquad \textit{energy density},$$
$$p^* = \frac{1}{d-1}(\mathrm{tr}(T) + \rho), \qquad\qquad\qquad \textit{(effective) pressure},$$
$$q = -T(V, \,\cdot\,) - \rho V^\flat, \qquad\qquad\qquad\qquad \textit{heat flux},$$
$$\Pi = T - q \vee V^\flat - pP - \rho V^\flat \otimes V^\flat, \qquad \textit{anisotropic pressure}.$$

Observe that V is an eigenvector field of T if and only if the heat flux vanishes. If bulk viscosity or similar effects are not explicitly considered, we do not distinguish between effective pressure and fluid pressure, and write $p = p^*$. Also note that if the Einstein–Hilbert equation $\mathrm{Ric} - \frac{\kappa}{2}g = T$ holds, we have

$$\rho = \mathrm{Ric}(V,V) + \frac{\kappa}{2}, \tag{3.2}$$

$$p = \frac{1}{d-1}\left(\mathrm{Ric}(V,V) - \frac{d-3}{2}\kappa\right), \tag{3.3}$$

$$q = -\mathrm{Ric}(V,V)V^\flat - \mathrm{Ric}(V, \,\cdot\,), \tag{3.4}$$

$$\Pi = \mathrm{Ric} + \frac{1}{d-1}\left(((d-2)\mathrm{Ric}(V,V) - \kappa)\,V^\flat \otimes V^\flat - (\mathrm{Ric}(V,V) + \kappa)g\right)$$
$$+ \mathrm{Ric}(V, \,\cdot\,) \vee V^\flat. \tag{3.5}$$

The function $\mathrm{Ric}(V,V)$ is also called the *Raychaudhuri scalar* (with respect to V).

Now, if T describes the energy–momentum of a perfect fluid with flow vector field V, we have $T = \rho V^\flat \otimes V^\flat + pP$, i.e., heat flux and anisotropic pressure vanish with respect to V, see Definition 5.0.3. If we evaluate the decomposition with respect to a different, future-directed reference frame W, one may compute in a straightforward manner:

$$\rho_W = (\rho + p)\gamma^2 - p, \quad p_W = p + \frac{1}{d-1}(\rho_W - \rho),$$

$$|q_W|_P^2 = (\rho + p)^2 \gamma^2 (\gamma^2 - 1), \quad |\Pi_W|_P^2 = \frac{d-2}{d-1}(\rho + p)^2 (\gamma^2 - 1)^2,$$

where $\gamma = g(V, W)$.

By the reverse Cauchy–Schwarz inequality it holds that $\gamma^2 \geq 1$ with equality if and only if $W = V$ at that point. Thus, if the *null energy condition* $\rho + p \geq 0$ holds, we have $\rho_W \geq \rho$, $p_W \geq p$, $|q_W|_P \geq 0$ and $|\Pi_W|_P \geq 0$ where equality holds only at points where $\rho + p = 0$ or $V = W$. Therefore, at points where the spacetime does not satisfy the Einstein condition $\mathrm{Ric} = k_2 g$ (see Definition 2.2.33), the flow direction of a perfect fluid is uniquely determined by its energy–momentum tensor.

Conversely, any symmetric tensor field T defined on a spacetime is uniquely determined by its values $T(W, W)$ ranging over all local reference frames W (see [SW77, p. 77]).

Remark 3.1.5. We would like to interpret the quantity $\gamma = g(V, W)$ in the above example physically. In the Lorentzian plane, we have the notion of (oriented) angles between future-pointing timelike vectors [BN84]. Similarly, let $E = W + g(V, W)V$ be the projection of W onto the restspace of V. Let $p \in M$. We may interpret E_p as the direction in which an instantaneous observer (p, V_p) would see a particle travel that were to move along an integral curve of W through p. The vectors V_p and E_p span a Lorentzian plane in $T_p M$, and we may define the angle $\phi(p)$ from V_p to W_p as the oriented angle with respect to the basis (V_p, E_p). This angle may be interpreted as the rapidity of the particle (as measured by (p, V_p)), whereas $\tanh(\phi(p))$ is the Newtonian velocity. By construction, we always have $\phi(p) \geq 0$ so that it is actually not an oriented angle: We simply have $g(V, W) = -\cosh(\phi)$ with $\phi \geq 0$. Of course, it still makes sense to say that another particle would have a negative velocity if it were travel in the direction $-E_p$.

3.2 Kinematical Quantities: Shear, Vorticity, Expansion and Acceleration

The $(D + 1)$-decomposition with respect to an observer field may be applied to its own covariant differential. This procedure yields the kinematical quantities (also called kinematical invariants) which describe the deformation of small spatial volumes along the integral curves of the reference frame.

Definition 3.2.1. The $(D+1)$-decomposition of the reference frame's covariant differential is given by

$$\nabla V^b = \sigma - \omega + \frac{\Theta}{d-1}P + \varepsilon V^b \otimes \dot{V}^b,$$

with the *kinematical quantities*:

$$
\begin{array}{lll}
\Theta = \operatorname{div} V, & \textit{expansion scalar}, & (3.6) \\[4pt]
\dot{V} = \nabla_V V, & \textit{acceleration}, & (3.7) \\[4pt]
\omega = -\dfrac{1}{2}\left(\mathrm{d}V^b + \varepsilon \dot{V}^b \wedge V^b\right), & \textit{vorticity tensor}, & (3.8) \\[8pt]
\sigma = \operatorname{sym}(\nabla V^b) - \dfrac{1}{2}\varepsilon \dot{V}^b \vee V^b - \dfrac{\Theta}{d-1}P & & (3.9) \\[8pt]
 = \dfrac{1}{2}\mathcal{L}_V g - \dfrac{1}{2}\varepsilon \dot{V}^b \vee V^b - \dfrac{\Theta}{d-1}P, & \textit{shear tensor}. & (3.10)
\end{array}
$$

We also define the *red-shift one-form* $\ell := -\varepsilon \dot{V}^b + \varepsilon \frac{\Theta}{d-1}V^b$.

Remark 3.2.2. For the expression (3.10), we used the well-known formula

$$(\mathcal{L}_Z g)(X,Y) = g(\nabla_X Z, Y) + g(\nabla_Y Z, X) = 2\operatorname{sym}(\nabla Z^b)(X,Y),$$

which holds for any vector fields X, Y, Z. In local coordinates, the kinematical quantities are given by

$$\Theta = \nabla_k V^k, \quad \dot{V}^i = V^k \nabla_k V^i, \quad \omega_{ij} = -P^k{}_i P^l{}_j \nabla_{[k}V_{l]},$$

$$\sigma_{ij} = P^k{}_i P^l{}_j \nabla_{(k}V_{l)} - \frac{\Theta}{d-1}P_{ij}.$$

Example 3.2.3. Let (M, g) be a Lorentzian manifold that admits a spin structure. For details on semi-Riemannian spin geometry, we refer to [Bau81, BL04]. For this example, we only need the following facts: Associated with the spin structure is a $\mathbb{C}$-vector bundle, the so-called *spinor bundle* ΣM. Furthermore, we have the *Clifford multiplication* $\Gamma(TM) \times \Gamma(\Sigma M) \to \Gamma(\Sigma M)$, $(X, \psi) \mapsto X \cdot \psi$ with the characteristic property $X \cdot Y \cdot \psi + Y \cdot X \cdot \psi = -2g(X,Y)\psi$, an indefinite Hermitian tensor bundle metric $\langle \cdot, \cdot \rangle$ over ΣM, and a *spin connection* $\nabla \colon \Gamma(TM) \times \Gamma(\Sigma M) \to \Gamma(\Sigma M)$ compatible with this metric. With this spin connection, the *Dirac operator* $\mathrm{D} \colon \Gamma(\Sigma M) \to \Gamma(\Sigma M)$ is defined by $\mathrm{D}\psi = \sum_{k=0}^{d-1} \varepsilon_i E_i \cdot \nabla_{E_i}\psi$ where $(E_i)_{i=0,\dots,d-1}$ is a local orthonormal frame and $\varepsilon_i = -1$ for $i = 0$ and $\varepsilon_i = 1$ for $i \in \{1, \dots, d-1\}$.

Let $\psi \in \Gamma(\Sigma M)$ be a spinor field such that its *Dirac current* $J_\psi \in \Gamma(TM)$ defined by $g(J_\psi, X) = -\langle X \cdot \psi, \psi \rangle$ is timelike. Since (in the Lorentzian case) the Dirac current of any spinor is either timelike or lightlike at any point, this may be seen as a generic case. Then, we may write $J = J_\psi = fV$ for a nowhere vanishing function $f \in \mathfrak{F}(M)$

and an observer field $V \in \Gamma(TM)$. The bilinear form $(\,\cdot\,,\,\cdot\,)$ over ΣM defined by $(\psi,\chi) := \langle V \cdot \psi, \chi \rangle$ is then a positive-definite Hermitian metric. We necessarily have $f = (\psi,\psi)$:

$$-f = g(V,V)f = g(J,V) = -\langle V \cdot \psi, \psi \rangle = -(\psi,\psi).$$

Observe that this shows that J_ψ is always future-pointing, and that ψ does not vanish anywhere since $(\,\cdot\,,\,\cdot\,)$ is positive-definite. We wish to compute the kinematical properties of the observer field V $(X, Y \in \Gamma(TM)$; "$\Re$" denotes the real part):

$$(\psi,\psi)\Theta = -2\Re\langle\psi, V \cdot \nabla_V \psi + \mathrm{D}\psi\rangle$$
$$(\psi,\psi)\dot{V}^\flat(X) = -2\Re\langle\psi, (g(V,X)V + X) \cdot \nabla_V \psi\rangle$$
$$(\psi,\psi)\omega(X,Y) = \Re\langle\psi, (g(V,Y)V + Y) \cdot \nabla_X \psi - (g(V,X)V + X) \cdot \nabla_Y \psi\rangle$$
$$+ \Re\langle\psi, (g(V,X)Y - g(V,Y)X) \cdot \nabla_V \psi\rangle$$
$$(\psi,\psi)\sigma(X,Y) = -\Re\langle\psi, (g(V,Y)V + Y) \cdot \nabla_X \psi + (g(V,X)V + X) \cdot \nabla_Y \psi\rangle$$
$$- \tfrac{2}{d-1}\Re\langle\psi, ((d-2)g(V,X)g(V,Y) - g(X,Y)) V \cdot \nabla_V \psi\rangle$$
$$- \Re\langle\psi, (g(V,Y)X + g(V,X)Y) \cdot \nabla_V \psi\rangle$$
$$+ \tfrac{2}{d-1}\Re\langle\psi, (g(V,X)g(V,Y) + g(X,Y)) \mathrm{D}\psi\rangle.$$

These formulas may be determined as follows. First, we have for arbitrary vector fields X, Y:

$$X(f) = X(\langle V \cdot \psi, \psi \rangle) = \langle \nabla_X(V \cdot \psi), \psi \rangle + \langle V \cdot \psi, \nabla_X \psi \rangle$$
$$= \langle \nabla_X V \cdot \psi, \psi \rangle + \langle V \cdot \nabla_X \psi, \psi \rangle + \langle V \cdot \psi, \nabla_X \psi \rangle$$
$$= -f \underbrace{g(V, \nabla_X V)}_{=0} + \overline{\langle\psi, V \cdot \nabla_X \psi\rangle} + \langle\psi, V \cdot \nabla_X \psi\rangle$$
$$= 2\Re\langle\psi, V \cdot \nabla_X \psi\rangle \; (= 2(\psi, \nabla_X \psi)), \tag{3.11}$$

and

$$g(\nabla_X J, Y) = X(g(J,Y)) - g(J, \nabla_X Y) = -X(\langle Y \cdot \psi, \psi \rangle) + \langle \nabla_X Y \cdot \psi, \psi \rangle$$
$$= -\langle \nabla_X(Y \cdot \psi), \psi \rangle - \langle Y \cdot \psi, \nabla_X \psi \rangle + \langle \nabla_X Y \cdot \psi, \psi \rangle$$
$$= -\langle \nabla_X Y \cdot \psi, \psi \rangle - \langle Y \cdot \nabla_X \psi, \psi \rangle - \langle Y \cdot \psi, \nabla_X \psi \rangle + \langle \nabla_X Y \cdot \psi, \psi \rangle$$
$$= -2\Re\langle\psi, Y \cdot \nabla_X \psi\rangle.$$

Then we may compute the divergence of J as follows:

$$\mathrm{div}\, J = \sum_{k=0}^{d-1} \varepsilon_i g(\nabla_{E_i} J, E_i) = \sum_{k=0}^{d-1} \varepsilon_i \left(-2\Re\langle\psi, E_i \cdot \nabla_{E_i} \psi\rangle\right) = -2\Re\langle\psi, \mathrm{D}\psi\rangle. \tag{3.12}$$

Combining Eqs. (3.11), (3.12) yields:

$$f\Theta = f\operatorname{div} V = f\operatorname{div}\left(f^{-1}J\right) = -V(f) + \operatorname{div} J = -2\Re\langle\psi, V\cdot\nabla_V\psi + \mathrm{D}\psi\rangle.$$

The acceleration is given as follows:

$$
\begin{aligned}
fg(\nabla_V V, X) = fg(\nabla_V\left(f^{-1}J\right), X) &= -V(f)g(V, X) + g(\nabla_V J, X)\\
&= -2\Re\langle\psi, (g(V, X)V + X)\cdot\nabla_V\psi\rangle.
\end{aligned}
$$

The shear and vorticity tensors follow from similar calculations.

Definition 3.2.4. We say that V is

1. *irrotational* if $\omega = 0$,

2. *shear-free* if $\sigma = 0$,

3. *rigid* if $\sigma = 0$ and $\Theta = 0$.

Remark 3.2.5. Obviously, V is geodesic if and only if the acceleration vanishes.

Proposition 3.2.6. Raychaudhuri's equation. *The following holds:*

$$\operatorname{Ric}(V, V) = -V(\Theta) - \frac{1}{d-1}\Theta^2 - |\sigma|_P^2 + |\omega|_P^2 + \operatorname{div}(\nabla_V V) \tag{3.13}$$

$$= -\operatorname{div}(\varepsilon(d-1)\ell^\sharp - (d-2)\nabla_V V) - \frac{d}{d-1}\Theta^2 - |\sigma|_P^2 + |\omega|_P^2. \tag{3.14}$$

Proof. This is a well-established identity. For a survey, also including spaces with non-vanishing torsion, see [KS07]. □

Remark 3.2.7. The second version (Eq. (3.14)) is of particular interest when one wishes to apply *Bochner's technique*: for compact manifolds, the divergence term vanishes via Gauss' theorem when integrating:

$$\int_M |\omega|^2\,\mathrm{dvol}(g) = \int_M \left(\operatorname{Ric}(V, V) + \tfrac{d}{d-1}\Theta^2 + |\sigma|^2\right)\mathrm{dvol}(g).$$

This immediately shows, for example, that compact Riemannian or Lorentzian manifolds with negative Ricci curvature do not admit rigid reference frames. This is related to the classical result that compact Riemannian manifolds with negative Ricci curvature do not admit non-trivial Killing vector fields [Boc46]. For Bochner's technique in Lorentzian geometry, see [RS98].

3.3 Shear-free Reference Frames

This section contains a collection of propositions which—more or less loosely—relate kinematics and conformal or causal structure, with particular emphasis on shear-free reference frames.

Proposition 3.3.1. *Let $\varphi \in \mathfrak{F}(M)$ and write $a = (\nabla_V V)^\flat$ for convenience. Under the conformal change $\tilde{g} = e^{2\phi}g$, $\tilde{V} = e^{-\phi}V$, the kinematical quantities transform as follows:*

$$e^\phi \tilde{\Theta} = \Theta + (d-1)d\phi(V), \quad \tilde{a} = a - \varepsilon P(\nabla\phi, \,\cdot\,),$$
$$e^{-\phi}\tilde{\omega} = \omega, \quad e^{-\phi}\tilde{\sigma} = \sigma, \quad \tilde{\ell} = \ell + d\phi.$$

Proof. We use the formulas stated in Proposition 2.3.4 to calculate the conformally transformed expressions of the kinematical invariants (3.6–3.10):

1. The expansion is computed from the divergence of the observer field:

$$e^\phi \tilde{\Theta} = e^\phi \widetilde{\mathrm{div}}\tilde{V} = e^\phi \left(-e^{-\phi}d\phi(V) + e^{-\phi}\widetilde{\mathrm{div}}V\right)$$
$$= -d\phi(V) + \mathrm{div}\,V + d \cdot d\phi(V) = \Theta + (d-1)d\phi(V).$$

2. Acceleration:

$$e^{2\phi}\tilde{\nabla}_{\tilde{V}}\tilde{V} = e^{2\phi}\tilde{\nabla}_{e^{-\phi}V}\left(e^{-\phi}V\right) = e^\phi \tilde{\nabla}_V \left(e^{-\phi}V\right) = -d\phi(V)V + \tilde{\nabla}_V V$$
$$= -d\phi(V)V + \nabla_V V + 2d\phi(V)V - g(V,V)\nabla\phi$$
$$= \nabla_V V + d\phi(V)V - \varepsilon\nabla\phi,$$

thus $\tilde{a} = \tilde{g}(\tilde{\nabla}_{\tilde{V}}\tilde{V}, \,\cdot\,) = a + d\phi(V)g(V, \,\cdot\,) - \varepsilon d\phi.$

3. Let $u = g(V, \,\cdot\,)$ and $\tilde{u} = \tilde{g}(\tilde{V}, \,\cdot\,) = e^\phi u$. Then we compute for the conformally transformed vorticity tensor:

$$2e^{-\phi}\tilde{\omega} = e^{-\phi}\left(-d\tilde{u} - \varepsilon\tilde{a} \wedge \tilde{u}\right) = -d\phi \wedge u - du + \left(-\varepsilon a - \varepsilon u(\nabla\phi)u + d\phi\right) \wedge u$$
$$= -du - \varepsilon a \wedge u = 2\omega.$$

4. Shear tensor:

$$e^{-\phi}\mathrm{sym}(\tilde{\nabla}\tilde{u}) = e^{-\phi}\left(\mathrm{sym}(\nabla\tilde{u}) - \tilde{u} \vee d\phi + \tilde{u}(\nabla\phi)g\right)$$
$$= \mathrm{sym}(d\phi \otimes u + \nabla u) - u \vee d\phi + u(\nabla\phi)g$$
$$= \mathrm{sym}(\nabla u) - \frac{1}{2}u \vee d\phi + u(\nabla\phi)g,$$

thus

$$
\begin{aligned}
\mathrm{e}^{-\phi}\tilde{\sigma} &= \mathrm{e}^{-\phi}\left(\mathrm{sym}(\tilde{\nabla}\tilde{u}) - \frac{\varepsilon}{2}\tilde{a} \vee \tilde{u} - \frac{\tilde{\Theta}}{d-1}\tilde{P}\right) \\
&= \mathrm{sym}(\nabla u) - \frac{1}{2}u \vee \mathrm{d}\phi + u(\nabla\phi)g - \frac{\varepsilon}{2}\left(a - \varepsilon\mathrm{d}\phi + u(\nabla\phi)u\right) \vee u \\
&\quad - \frac{\Theta + (d-1)u(\nabla\phi)}{d-1}(g - \varepsilon u \otimes u) \\
&= \mathrm{sym}(\nabla u) - \frac{\varepsilon}{2}a \vee u - \frac{\Theta}{d-1}(g - \varepsilon u \otimes u) = \sigma.
\end{aligned}
$$

Finally,

$$
\tilde{\ell} = -\varepsilon\tilde{a} + \varepsilon\frac{\tilde{\Theta}}{d-1}\tilde{u} = -\varepsilon a - \varepsilon\mathrm{d}\phi(V)u + \mathrm{d}\phi + \varepsilon\frac{\Theta + (d-1)\mathrm{d}\phi(V)}{d-1}u = \ell + \mathrm{d}\phi.
$$

$\square$

The above proposition shows that the vorticity and shear tensors as well as the exterior derivative of the red-shift form are conformal invariants. These quantities are therefore expected to be tied to the conformal and causal structure of (M, g), which we elaborate in the following.

Remark 3.3.2. The red-shift form $\ell(X) = -\varepsilon g(\nabla_V V, X) + \varepsilon\theta g(V, X)$, setting $\theta := \frac{\Theta}{d-1}$ for convenience, behaves under conformal transformations precisely like a one-form generating a Weyl connection. Indeed, this Weyl connection,

$$
\mathrm{D}_X Y := \nabla_X Y - \ell(X)Y - \ell(Y)X + g(X, Y)\ell^\sharp,
$$

has some interesting properties:

- For any $X, Y \in \Gamma(TM)$, it holds that

$$
V(g(X, Y)) = g(\mathrm{D}_V X, Y) + g(X, \mathrm{D}_V Y) + 2\theta g(X, Y).
$$

Thus, parallel transport with respect to this connection along integral curves of V maps horizontal orthogonal frames onto horizontal orthogonal frames, with possible rescaling. Indeed, for vanishing expansion, it coincides with the Fermi–Walker transport along those curves.

- The general formula for calculating the Ricci curvature of a Weyl connection with generating one-form ℓ is given by (see for example [Nar01]):

$$
\begin{aligned}
\mathrm{Ric}^{\mathrm{D}}(X, Y) &= \mathrm{Ric}^{\nabla}(X, Y) + (d-1)(\nabla_Y \ell)(X) - (\nabla_X \ell)(Y) \\
&\quad + \mathrm{div}(\ell^\sharp)g(X, Y) - (d-2)g(\ell^\sharp, \ell^\sharp) + (d-2)\ell(X)\ell(Y),
\end{aligned}
$$

where ∇ is the Levi-Civita derivative. If we insert the red-shift form ℓ and set $X = Y = V$, we have:

$$\mathrm{Ric}^{\mathrm{D}}(V, V) = \mathrm{Ric}^{\nabla}(V, V) + (d - 2)\left((\nabla_V \ell)(V) + \ell(V)^2 - \varepsilon g(\ell^\sharp, \ell^\sharp)\right) + \varepsilon \operatorname{div}(\ell^\sharp)$$
$$= \mathrm{Ric}^{\nabla}(V, V) + (d - 1)V(\theta) - \operatorname{div}(\nabla_V V) + (d - 1)\theta^2.$$

Eliminating $\mathrm{Ric}^{\nabla}(V, V)$ via the usual Raychaudhuri equation (Eq. (3.13)), we arrive at a remarkable conformal version of said equation:

$$\mathrm{Ric}^{\mathrm{D}}(V, V) = |\omega|_P^2 - |\sigma|_P^2.$$

A deeper significance to this formula remains unclear, however.

Definition 3.3.3. The reference frame V is *synchronizable* if there exist functions $\beta, t \in \mathfrak{F}(M)$ with $V = -\beta \nabla t$. It is called *proper time synchronizable* if there exists a function $t \in \mathfrak{F}(M)$ with $V = -\nabla t$.

Remark 3.3.4. The terminology is usually reserved for spacetimes but we may apply it to the Riemannian case as well. From Theorem 2.3.19 it is obvious that a spacetime is stably causal if and only if it admits a synchronizable reference frame.

Proposition 3.3.5. *The following holds:*

1. *Any synchronizable reference frame is irrotational.*

2. *Any irrotational reference frame is locally synchronizable.*

3. *Any proper time synchronizable reference frame is irrotational and geodesic.*

4. *If the first Betti number vanishes, any irrotational and geodesic reference frame is proper time synchronizable.*

Proof. See [O'N83, pp. 358–360]. Concering statement (4), the condition that the manifold M is simply connected is easily generalized to $B_1(M) = 0$. $\square$

Remark 3.3.6. Although $\omega = 0$ implies that *locally* one has $V = -\beta \nabla t$ there is no obvious topological condition for this to hold globally – see [OS90] for a counter-example in $\mathbb{R}^4$. The reason for this is that statement (4) is implied by de Rham cohomology. Item (2) on the other hand rests on Frobenius' theorem. Although local integral submanifolds may be glued together to define a foliation, the leaves of this foliation might not be embedded submanifolds [Ste64, p. 136].

Definition 3.3.7. A vector field $K \in \Gamma(TM)$ is called a *conformal vector field* if $\mathcal{L}_K g = 2\phi g$ for some function $\phi \in \mathfrak{F}(M)$. If $\phi = 0$, K is a *Killing vector field*.

We call V a *conformally stationary* reference frame if V is proportional to a conformal vector field. We say that V is *stationary* if it is proportional to a Killing vector field. If V is stationary and irrotational, it is *static*.

Remark 3.3.8. In local coordinates, $K \in \Gamma(TM)$ is conformal if and only if $\nabla_i K_j + \nabla_j K_i = 2\phi g_{ij}$ holds. See [HS91] for a discussion of conformal vector fields in four-dimensional spacetimes.

Proposition 3.3.9. *A vector field is conformal if and only if its local flow is a one-parameter group of local conformal transformations. It is Killing if and only if its local flow is a one-parameter group of local isometries.*

Proof. For a proof in coordinates, see [DS99, p. 49]. We follow the argument in [O'N83, p. 251] where the Killing case is discussed. Let K be a vector field such that its local flow ψ_t, $t \in \,]-\epsilon, \epsilon[$, is a one-parameter family of local conformal transformations: $\psi_t^* g = e^{2\phi_t} g$ for some family of local functions ϕ_t. It is easy to see that $\psi_t \circ \psi_s = \psi_{t+s}$ and $\psi_0 = \mathrm{id}$ implies that this family is of the form $\phi_t = t\phi$. The Lie derivative of g in the direction of K thus computes to

$$\mathcal{L}_K g = \lim_{t \to 0} \frac{1}{t}(\psi_t^* g - g) = \lim_{t \to 0} \frac{1}{t}(e^{2t\phi} - 1)g = 2\phi g.$$

Conversely, let $\mathcal{L}_K g = 2\phi g$. For any tangent vector v at a point in the domain of the flow, $w = \psi_{s*}(v)$ is a tangent vector in the domain for small s, and

$$2\phi g(w, w) = \lim_{t \to 0} \frac{1}{t}\left(g(\psi_{t*}(w), \psi_{t*}(w)) - g(w, w)\right),$$

or

$$2\phi g(\psi_{s*}(v), \psi_{s*}(v)) = \lim_{t \to 0} \frac{1}{t}\left(g(\psi_{s+t*}(v), \psi_{s+t*}(v)) - g(\psi_{s*}(v), \psi_{s*}(v))\right).$$

Thus, the function $s \mapsto g(\psi_{s*}(v), \psi_{s*}(v))$ has derivative $2\phi g(\psi_{s*}(v), \psi_{s*}(v))$ which implies $g(\psi_{s*}(v), \psi_{s*}(v)) = e^{2s\phi} g(v, v)$. □

Definition 3.3.10. Cf. [GRK96]. A cospacelike vector field $K \in \Gamma(TM)$ is called *spatially conformal* if for some function $\phi \in \mathfrak{F}(M)$, $(\mathcal{L}_K g)(X, Y) = 2\phi g(X, Y)$ holds for all vector fields X, Y perpendicular to K.

Proposition 3.3.11. *Let $K \in \Gamma(TM)$ be a cospacelike vector field and $V = \dfrac{K}{\sqrt{|g(K,K)|}}$. Then, the following statements are equivalent:*

1. *The vector field K is spatially conformal.*

2. *The vector field fK is spatially conformal for any function $f \in \mathfrak{F}(M)$.*

3. *It holds that $\mathcal{L}_V P = \frac{2\Theta}{d-1} P$.*

4. *The reference frame V is shear-free.*

5. *The local flow ψ_t of V is a one-parameter group of local conformal transformations with respect to the transversal vector bundle metric P.*

Proof. Obviously, (2) $\Rightarrow$ (1). Now observe that for any function $f \in \mathfrak{F}(M)$:

$$\mathcal{L}_{fV}g = f\mathcal{L}_V g + \mathrm{d}f \vee V^\flat = 2f\sigma + \varepsilon f \dot{V}^\flat \vee V^\flat + \frac{2f\Theta}{d-1}P + \mathrm{d}f \vee V^\flat. \qquad (3.15)$$

If $K = fV$ is spatially conformal, this means that for all X, Y perpendicular to V:

$$2f\sigma(X,Y) + \frac{2f\Theta}{d-1}P(X,Y) = (\mathcal{L}_{fV}g)(X,Y) = 2\phi g(X,Y) = 2\phi P(X,Y).$$

Taking the trace yields $\phi = \frac{f\Theta}{d-1}$ and $\sigma = 0$, proving (1) $\Rightarrow$ (4). To show (4) $\Rightarrow$ (3), with the help of Eq. (3.10) we compute for $\sigma = 0$:

$$\mathcal{L}_V P = \mathcal{L}_V(g - \varepsilon V^\flat \otimes V^\flat) = 2\sigma + \varepsilon \dot{V}^\flat \vee V^\flat + \frac{2\Theta}{d-1}P - \varepsilon(\mathcal{L}_V V^\flat) \vee V^\flat$$

$$= \varepsilon \dot{V}^\flat \vee V^\flat + \frac{2\Theta}{d-1}P - \varepsilon \dot{V}^\flat \vee V^\flat = \frac{2\Theta}{d-1}P.$$

By similar arguments, (3) implies that V is shear-free which in turn implies that any multiple fV of V, and therefore any multiple of K, is spatially conformal:

$$(\mathcal{L}_{fV}g)(X,Y) = \frac{2f\Theta}{d-1}P(X,Y) = \frac{2f\Theta}{d-1}g(X,Y).$$

Finally, (5) is equivalent to (3): By similar arguments as in Proposition 3.3.9 we have $\psi_t^* P = \exp(\frac{\Theta}{d-1})P$, where $\psi = \psi_t$ is the local flow of V. Since V is invariant with respect to its own local flow ($\mathcal{L}_V V = [V, V] = 0$), for any $v \in H_p M$, we have $\psi_{*p}(v) \in H_{\psi(p)}M$ because:

$$P_{\psi(p)}(\psi_{*p}(v), V_{\psi(p)}) = P_{\psi(p)}(\psi_{*p}(v), \psi_{*p}(V_p)) = (\psi^* P)_p(v, V_p)$$

$$= \exp(\tfrac{\Theta(p)}{d-1})P(v, V_p) = 0.$$

This shows that ψ also preserves the fibers of HM. $\qquad\square$

Remark 3.3.12. We would like to note yet another characterization of a shear-free reference frame: The projections onto VM and HM define a semi-Riemannian almost product structure. Corollary 2.2 in [Mon83] shows that the horizontal distribution is umbilical if and only if V is shear-free.

Also observe that the vector field V generates a one-parameter family of local isometries with respect to the Riemannian metric P if and only if it is rigid. Therefore, one may refer to rigid reference frames as *spatially stationary*, cf. [RS98].

The calculations above show that the conformal factor of a spatially conformal vector field is given by $\phi = \frac{1}{d-1}|K|\operatorname{div}\frac{K}{|K|}$ with $|K| = \sqrt{|g(K,K)|}$. The spatially conformal vector field, and thereby this factor, may be scaled by any positive function. Thus, we may restrict the property of being spatially conformal to reference frames without loss of generality. In contrast to this, cospacelike conformal vector fields are unique up to a constant, as shown by the next proposition.

Proposition 3.3.13. *Suppose K and L are cospacelike conformal vector fields which are parallel, i.e., there exists a nowhere vanishing function $f \in \mathfrak{F}(M)$ such that $L = fK$. Then, f is locally constant.*

Proof. Let $\mathcal{L}_K g = 2\phi g$ and $\mathcal{L}_L g = 2\psi g$. Now,

$$2\psi g = \mathcal{L}_L g = \mathcal{L}_{fK} g = f\mathcal{L}_K g + \mathrm{d}f \vee K^\flat = 2f\phi g + \mathrm{d}f \vee K^\flat,$$

which implies

$$2\psi g(K, K) = 2f\phi g(K, K) + 2\mathrm{d}f(K)g(K, K),$$

or $\mathrm{d}f(K) = \psi - f\phi = 0$. If E is any local vector field perpendicular to K, we also have

$$\psi g(K, E) = 0 = \mathrm{d}f(E)g(K, K).$$

Thus, $\mathrm{d}f = 0$. $\qquad\square$

For $d \geq 3$, we may drop the assumption that the conformal vector fields are cospacelike:

Proposition 3.3.14. *Assume that M is connected, and $d \geq 3$. Suppose K and L are non-trivial conformal vector fields which are parallel, i.e., both K and L do not vanish identically, and there exists a function $f \in \mathfrak{F}(M)$ such that $L = fK$. Then, f is constant.*

Proof. First we note that conformal vector fields are uniquely determined by their value, and the value of their first and second covariant derivatives at one point. In another terminology, a conformal vector field is uniquely determined by its two-jet at one point [Der12]. This may be seen by adjusting the argument given in [Wal84, Section C.3], where it is proved that Killing vector fields are uniquely determined by their value and the value of their first covariant derivative at one point. The argument also shows that if a conformal vector field vanishes on an open subset $U \subset M$, it vanishes on all of M.

Now, as we have seen in the proof of Proposition 3.3.13,

$$2(\psi - f\phi)g = \mathrm{d}f \vee K^\flat \tag{3.16}$$

for some functions ϕ and ψ. Set up a local frame. The rank of the matrix on the left-hand side is either zero or $d \geq 3$. The rank of the matrix on the right-hand side is at most two. Therefore, the tensor fields on both sides of Eq. (3.16) must vanish. Since K does not vanish on any open subset by the above remarks, $\mathrm{d}f$ must vanish. Therefore, f is constant. $\qquad\square$

Remark 3.3.15. The above proposition is false for $d = 2$ if g is a Lorentzian metric. A counter-example is given by the flat metric $g = -\mathrm{d}t^2 + \mathrm{d}x^2$ and the conformal vector field $K = \partial_t + \partial_x$. For any function $\alpha \colon \mathbb{R} \to \mathbb{R}$, the vector field $L = \alpha(t+x)K$ is conformal, as well.

Obviously, any conformally stationary observer field is spatially conformal. The following proposition characterizes the cases where the converse statement is true as well.

Proposition 3.3.16. *Cf. [HP88]. Let V be a spatially conformal, i.e., shear-free, reference frame. There exists a positive function $f \in \mathfrak{F}(M)$ such that $K = fV$ is conformal if and only if the red-shift form ℓ of V is exact.*

Proof. By projecting (3.15) onto V and the subspace perpendicular to V one may infer that $\mathcal{L}_{fV} g = \frac{2f\Theta}{d-1} g$ implies

$$\mathrm{d}(\log f) = -\varepsilon \dot{V}^\flat + \varepsilon \frac{\Theta}{d-1} V^\flat = \ell.$$

Thus, ℓ is indeed exact if fV is conformal.

Conversely, suppose $\mathrm{d}(\log f) = -\varepsilon \dot{V}^\flat + \varepsilon \frac{\Theta}{d-1} V^\flat$ for some function f. Plugging this into (3.15) immediately shows that fV is conformal. □

Corollary 3.3.17. *The following holds:*

1. *A reference frame is locally conformally stationary if and only if it is shear-free and its red-shift form is closed.*

2. *A reference frame is locally stationary if and only if it is rigid and the curl of its acceleration vanishes.*

The statements hold globally if the first Betti number vanishes.

Proof. Apply Poincaré's lemma and de Rham cohomology to Proposition 3.3.16, and note that the conformal factor is proportional to the expansion (Proposition 3.3.11). □

Corollary 3.3.18. *Let V be conformally stationary. There exists a conformal change $\tilde{g} = \mathrm{e}^{2\phi} g$, $\tilde{V} = \mathrm{e}^{-\phi} V$ such that $\tilde{V}$ is stationary and geodesic with respect to $\tilde{g}$.*

Proof. In another terminology [KR97], cospacelike conformal vector fields are always *inessential.* This is quite well-known but may be rephrased in terms of kinematics: Write $a = (\nabla_V V)^\flat$ for convenience. The shear is conformally invariant, and we may choose $\phi = -\log f$ so that $\tilde{l} = 0$, thus $\tilde{\Theta} \tilde{u} = (d-1)\tilde{a}$ which implies $\tilde{\Theta} = 0$ and $\tilde{a} = 0$. □

Proposition 3.3.19. *Suppose (M, g) is Lorentzian, and V is conformally stationary. Then, (M, g) is stably causal if the inequality*

$$|\nabla_V V|_P < \frac{|\Theta|}{d-1} \tag{3.17}$$

holds. Conversely, if (M, g) is stably causal, there exists a conformal change $\tilde{g} = e^{2\phi}g$, $\tilde{V} = e^{-\phi}V$ such that

$$|\tilde{\nabla}_{\tilde{V}} \tilde{V}|_{\tilde{P}} < \frac{|\tilde{\Theta}|}{d-1}$$

holds.

Proof. If V is conformally stationary, there exists a function $f \in \mathfrak{F}(M)$ such that $\mathrm{d}(\log f) = \dot{V}^\flat - \frac{\Theta}{d-1}V^\flat$ holds. Now, f or $-f$ is a temporal function if and only if $g(\nabla f, \nabla f) < 0$ if and only if the stated inequality holds.

Conversely, let ϕ be a temporal function. Because of Corollary 3.3.18 we may first apply a conformal change such that the acceleration and expansion vanish: $\nabla_V V = 0$, $\Theta = 0$. Now applying the conformal change $\tilde{g} = e^{2\phi}g$, $\tilde{V} = e^{-\phi}V$ we see with Proposition 3.3.1 that

$$0 > g(\nabla \phi, \nabla \phi) = P(\nabla \phi, \nabla \phi) - \mathrm{d}\phi(V)^2 = e^{2\phi}\tilde{g}(\tilde{\nabla}_{\tilde{V}} \tilde{V}, \tilde{\nabla}_{\tilde{V}} \tilde{V}) - e^{2\phi}\frac{\tilde{\Theta}^2}{(d-1)^2}.$$

$\square$

Corollary 3.3.20. *Suppose M is compact, and V is conformally stationary. If g is Riemannian, there exists a point $p \in M$ with $|\dot{V}_p| = \Theta(p) = 0$; if g is Lorentzian, there exists a point $p \in M$ with $|\dot{V}_p| \geq |\Theta(p)|$.*

Proof. Since M is compact, f attains an extremum at some point $p \in M$, thus $\mathrm{d}f_p = 0$. The result easily follows in the Riemannian case. In the compact Lorentzian case, (M, g) is non-stably causal and (3.17) must be violated at some point (Proposition 2.3.17). $\square$

Example 3.3.21. We wish to extend Example 3.2.3. If ψ is a twistor spinor (i.e., $\nabla_X \psi + \frac{1}{d}X \cdot \mathrm{D}\psi = 0$ for all $X \in \Gamma(TM)$), it is a well-known fact that the corresponding Dirac current is a conformal vector field. Indeed, we have

$$\Theta = -\frac{2(d-1)}{d}(\psi, \psi)^{-1}\Re\langle\psi, \mathrm{D}\psi\rangle, \quad \sigma = 0,$$

$$\ell(X) = -\frac{2}{d}(\psi, \psi)^{-1}(\psi, X \cdot \psi) = X[\log[(\psi, \psi)]].$$

If ψ is a Killing spinor (i.e., $\nabla_X \psi + \lambda X \cdot \psi = 0$ for all $X \in \Gamma(TM)$ and some $\lambda \in \mathbb{C}$), we also have $\omega = 0, \dot{V} = 0$. A spinor is Killing if and only if it is twistor and and a solution to the Dirac equation $D\psi = d\lambda\psi$, $\lambda \in \mathbb{C}$. For a more detailed exposition including metrics of other signatures, see [Boh98].

The physically most relevant case of a Killing spinor is the Dirac field with mass $m \geq 0$ which satisfies $D\psi = -im\psi$, i.e., for purely imaginary λ. In this case, the Dirac current is a covariantly constant vector field (assuming that it is timelike).

Finally, we have the following:

Proposition 3.3.22. *Let (M, g) be a spacetime that admits a conformally stationary reference frame V. If (M, g) contains a closed timelike curve, it contains a closed timelike curve through every point, i.e., it is totally vicious.*

Proof. For the stationary case, see [CJ88] or [Jos93, p. 113]. But every conformally stationary reference frame is stationary with respect to a conformal metric (Corollary 3.3.18), which induces the same causal structure. □

Remark 3.3.23. Note that Propositions 2.3.17 and 3.3.22 imply Theorem 1.1 in [Sán06].

Chapter 4

TILTED TWISTED AND WARPED PRODUCTS

We have already mentioned that V induces an orthogonal splitting of the tangent bundle. This splitting only implies a local metric structure in twisted product form if V is shear-free, geodesic and irrotational. In this chapter, we investigate a product structure which applies to shear-free reference frames in general.

4.1 Basic Properties of Tilted Products

Locally, any metric admitting a shear-free reference frame can be brought into a standard form, see also [COS01, GPS+10]:

Proposition 4.1.1. *Let (M, g) be a Riemannian or Lorentzian manifold admitting a shear-free reference frame V, $u = g(V, \cdot)$. Then about each point $p \in M$ there exists a chart (U, ϕ) with coordinates $\phi = (x^0, x^1, \ldots, x^{d-1}) = (t, x)$ such that $V = \partial_t$ and*

$$g_{ij}(t, x) = \varepsilon u_i(t, x) u_j(t, x) + a^2(t, x) h_{ij}(x), \tag{4.1}$$

where $a > 0$ is a function on $\phi(U)$.

Proof. Since V has no zeros, there exist local coordinates $\phi = (t, x)$ such that $V = \partial_t$, hence $V^i = u^i = \delta^i{}_0$ [O'N83, p. 30]. We may assume without loss of generality that $\phi(U)$ is of the form $I \times U_0$ where $I \subset \mathbb{R}$ is an open interval with $0 \in I$. In this comoving coordinate system, $\varepsilon = g_{ij} u^i u^j = g_{00}$ and $u_i = g_{ki} u^k = g_{0i}$. The coordinates of the covariant derivative of u are thus given by

$$\nabla_i u_j = \partial_i g_{0j} - \frac{1}{2} g^{kl} g_{0k} (\partial_i g_{jl} + \partial_j g_{il} - \partial_l g_{ij})$$
$$= \frac{1}{2} (\partial_i g_{0j} - \partial_j g_{0i} + \partial_0 g_{ij}).$$

For the components of the acceleration, one has in particular:

$$\dot{u}_i = u^k \nabla_k u_i = \nabla_0 u_i = \partial_0 g_{0i} - \frac{1}{2} \partial_i g_{00} = \partial_0 g_{0i}.$$

In the same manner, we obtain a similar expression for the components of the shear tensor:

$$\sigma_{ij} = \frac{1}{2}(\partial_0 g_{ij} + \varepsilon g_{0j}\partial_0 g_{0i} + \varepsilon g_{0i}\partial_0 g_{0j}) - \frac{1}{d-1}\Theta P_{ij}$$
$$= \frac{1}{2}\partial_0 P_{ij} - \frac{1}{d-1}\Theta P_{ij}.$$

Now suppose that $\sigma_{ij} = 0$. We can solve the last equation for the components of the projection tensor P_{ij}, which leads to

$$P_{ij}(t, x) = a^2(t, x)P_{ij}(0, x). \tag{4.2}$$

with

$$a(t, x) = \exp\left(\frac{1}{d-1}\int_0^t \Theta(\tau, x)\,\mathrm{d}\tau\right). \tag{4.3}$$

Defining $h_{ij}(x) = P_{ij}(0, x)$ yields the result. $\qquad\square$

Setting $b_i := t_i - \varepsilon u_i$ in the local expression (4.1) for shear-free manifolds, $t_i := \delta^0{}_i$, we see that the observer field is g-perpendicular to b: $u^k b_k = \delta^k{}_0\delta^0{}_k - \varepsilon \cdot u^k u_k = 0$. This motivates the definition of the following geometric structure:

Definition 4.1.2. Let $\varepsilon \in \{-1, +1\}$. We define the *tilted twisted product* of an open interval $(I, \varepsilon \mathrm{d}t^2)$ with a connected Riemannian manifold (N, h) of dimension $D \geq 1$ to be the product manifold $M = I \times N$, furnished with the metric tensor

$$g = \varepsilon(\pi_1^*\mathrm{d}t - b)^2 + a^2\pi_2^*h, \tag{4.4}$$

where a is a given real-valued, positive function on M, and b is a horizontal one-form on M, i.e., $b(\hat{\partial}_t) = 0$, where ∂_t is the canonical vector field on I.

Recall that $\hat{\partial}_t$ is the lift of ∂_t (Definition 2.1.14). The *standard observer field* $V = \hat{\partial}_t$ is shear-free. Tilted twisted products may be seen as characterized by a one-parameter family of conformally equivalent metrics $h_t(x) = a^2(t, x)h(x)$ and one-forms $b_t(x) = b(t, x)$ defined on N. We call b the *shift form*.

We also write $(M, g) = (I, \varepsilon \mathrm{d}t^2) \times_a^b (N, h)$ for this structure. If the *twisting function* a only depends on t, we speak of a *tilted warped product*, and a is the *warping function* or *scale parameter*. Trivial shift forms ($b = 0$) or twisting functions ($a = 1$) are omitted notationally, i.e., we write $(I, \varepsilon \mathrm{d}t^2) \times_a (N, h) := (I, \varepsilon \mathrm{d}t^2) \times_a^0 (N, h)$ and $(I, \varepsilon \mathrm{d}t^2) \times^b (N, h) := (I, \varepsilon \mathrm{d}t^2) \times_1^b (N, h)$.

We call (N, h) the *fiber* and each hypersurface $N_s := \{s\} \times N$ a *slice*. Obviously, each slice is diffeomorphic to the fiber, but the "metric" induced on N_s by g may be Riemannian, Lorentzian or degenerate, or of changing signature.

The vector bundle metrics P and π_2^*h over HM are conformally related: $P = a^2\pi_2^*h$. We would also like to mention that while the flow of V is a one-parameter family of conformal transformations with respect to the vector bundle metric P (Proposition 3.3.11), the same flow is a one-parameter family of isometries with respect to π_2^*h. From now on we always use π_2^*h to compute the norm of horizontal tensors.

Also, to keep the notation clear, in the following we frequently omit the " $\hat{\ }$ "-notation for lifts, or the projections π_1, π_2.

Remark 4.1.3. The terminology refers to twisted products [PR93] and tilted cosmological models [CHL06]. Twisted products may be viewed upon as tilted twisted products with vanishing shift form b. Note, however, that tilted twisted products with exact shift form may be written as twisted products, as well – see Proposition 4.1.12 below. Warped products are twisted products where the twisting function a is constant on each slice, i.e., it only depends on t. A twisted product might still be written in warped product form even if a is not constant on each slice, cf. [FLGRKÜ01]. We also refer to [Che11, Chapter 4] for more details on twisted and warped products.

Remark 4.1.4. One might also call tilted twisted products "standard shear-free manifolds" analogous to the term "standard stationary" [JS08]. Note, however, that standard stationary metrics are usually assumed to be stably causal. Also, applying a conformal change to the metric (4.4) yields a shear-free model that is not necessarily globally isometric to a tilted twisted product.

Remark 4.1.5. Of course, one might also wish to investigate products of the form $(M, g) = (S^1, \varepsilon d\phi^2) \times_a^b (N, h)$. Furthermore, a possible generalization to base spaces of arbitrary dimension would be the following: Let (N_1, h_1) be a parallelizable semi-Riemannian manifold of dimension k, i.e., with trivial tangent bundle TN_1. Lie groups, for example, are parallelizable, as well as all compact orientable three-dimensional manifolds [FV04, Section 11.3]. This implies that there exist globally defined one-forms $E^1, \ldots, E^k$ (with dual basis $E_1, \ldots, E_k$) such that $h_1 = \varepsilon_1(E^1)^2 + \cdots + \varepsilon_k(E^k)^2$, $\varepsilon_i = \pm 1$. Given the product $M = N_1 \times N$ and transversal shift forms $b^1, \ldots, b^k$ (i.e., $b^i(E_j) = 0$ for all $i, j = 1, \ldots, k$) one may then write down the metric tensor

$$g = \sum_{l=1}^{k} \varepsilon_l(\pi_1^*E^l - b^l)^2 + a^2\pi_2^*h.$$

We do not further discuss these possibilities in this work.

Proposition 4.1.6. Let $(M, g) = (I, \varepsilon dt^2) \times_a^b (N, h)$ be a tilted product. If $\varepsilon = 1$, (M, g) is a connected Riemannian manifold; if $\varepsilon = -1$, (M, g) is a connected Lorentzian manifold.

Proof. This is rather obvious by construction but we wish to provide an independent proof. Denote by $g^{(\pm)}$ the metric corresponding to $\varepsilon = \pm 1$. The tensor $g^{(+)}$ is a Riemannian metric: Any non-zero tangent vector v at $p \in M$ may be written as the linear combination $v = \alpha \, \partial_t|_p + \beta w$ with $\alpha^2 + \beta^2 > 0$, where w is tangent to the slice through p. It may be readily seen that

$$g_p^{(+)}(v, v) = (\alpha - \beta b_p(w))^2 + \beta^2 (a(p))^2 h_p(w, w) > 0,$$

and $g^{(+)}(V, V) = 1$ in particular. As for $g^{(-)}$, observe that $g^{(-)} = -2g^{(+)}(V, \cdot) \otimes g^{(+)}(V, \cdot) + g^{(+)}$. Complete V to a $g^{(+)}$-orthonormal frame $(V, E_1, \ldots, E_{d-1})$; then $g^{(-)}(E_i, E_j) = \delta_{ij}$, $g^{(-)}(V, E_i) = g^{(+)}(V, E_i) = 0$ but $g^{(-)}(V, V) = -1$. $\quad\square$

Remark 4.1.7. Note that the mapping $g^{(-)} \mapsto g^{(+)}$ may formally be thought of as a kind of Wick rotation [Wic54]: "$g^{(-)}(\sqrt{-1} \cdot V, \sqrt{-1} \cdot V) = g^{(+)}(V, V)$".

In order to analyze the causal structure and calculate the Ricci curvature of tilted warped products, we need the following (cf. [GKS05]):

Lemma 4.1.8. *The inverse metric of a tilted twisted product $(M, g) = (I, \varepsilon \mathrm{d}t^2) \times_a^b (N, h)$ is given by*

$$g^{-1} = \frac{1}{a^2} \left(h^{-1} + (|b|^2 + \varepsilon a^2) \partial_t \otimes \partial_t + \partial_t \vee (h^{-1}(b, \cdot)) \right),$$

where $|b| = |b|_h$.

Proof. Given adapted coordinates on $I \times N$, the metric tensor has the following partitioned matrix form:

$$(g_{ij}) = \begin{pmatrix} \varepsilon & -\varepsilon b \\ -\varepsilon b^T & a^2 h + \varepsilon b^T b \end{pmatrix}.$$

It is a simple matter to check that

$$(g^{ij}) = \frac{1}{a^2} \begin{pmatrix} b h^{-1} b^T + \varepsilon a^2 & b h^{-1} \\ h^{-1} b^T & h^{-1} \end{pmatrix}$$

is its inverse. $\quad\square$

Remark 4.1.9. The above formula may be derived from general formulas for the matrix inverse of partitioned matrices, see [HS81]. In adapted coordinates $(t, x) = (t, x_1, \ldots, x_{d-1})$, we have

$$g^{ij} = \frac{1}{a^2} \left(\bar{h}^{ij} + (|b|^2 + \varepsilon a^2) u^i u^j + u^i \bar{b}^j + u^j \bar{b}^i \right) \tag{4.5}$$

with $\bar{h}^{ik} h_{kj} = \delta^i{}_j - \delta^0{}_j \delta^i{}_0 = \delta^i{}_j - t_j u^i$ and $\bar{b}^j = \bar{h}^{jk} b_k = -\varepsilon \bar{h}^{jk} u_k$. Any time an index is raised with $\bar{h}^{ij}$, we indicate this with an overbar.

4.1.1 Examples of Tilted Products. Conformal Transformations, Changes of Slicing.

Example 4.1.10. Stationary spacetimes are important examples of shear-free spacetimes. Locally, any stationary spacetime (M', g') is of the form $M' = I' \times N$, with metric

$$g' = -\beta^2(\mathrm{d}t' + \xi)^2 + h,$$

where $\beta > 0$ is a function, ξ a one-form, and h a Riemannian metric on N. We then have that the reference frame $V = -\beta(\mathrm{d}t' + \xi)^\sharp$ is stationary, and in particular, shear-free. In fact, (M', g') is isometric to $(M, g) = (I \times N, g)$ with

$$g = -(\mathrm{d}t - t\mathrm{d}(\log \beta) + \beta\xi)^2 + h.$$

The isometry is given by $(t, x) = (t'\beta(x), x)$, as can be easily seen by a direct computation:

$$\mathrm{d}t - t\mathrm{d}(\log \beta) + \beta\xi = \mathrm{d}t'\beta + t'\mathrm{d}\beta - t'\beta\mathrm{d}(\log \beta) + \beta\xi = \beta(\mathrm{d}t' + \xi).$$

Thus, stationary spacetimes may locally be written in tilted product form.

Example 4.1.11. Suppose (M, g) is a Gödel-type spacetime as defined in [CS00, CSS02], i.e., we have $M = \mathbb{R}^2 \times N_0$ and

$$g = -C\mathrm{d}t'^2 + B\mathrm{d}t' \vee \mathrm{d}y + A\mathrm{d}y^2 + h_0$$

where (t', y) are the standard coordinates on $\mathbb{R}^2$, h_0 is a Riemannian metric on N_0, and $A, B, C \in \mathfrak{F}(N_0)$ are functions with $B^2 + AC > 0$. A particular Gödel-type spacetime may not be time-orientable and therefore cannot be a tilted product in such cases. However, if $C > 0$ and $A + B > 0$, the transformation $t = \sqrt{C}t'$ shows that (M, g) is a tilted product:

$$g = -(\mathrm{d}t - \frac{t}{2}\mathrm{d}(\log C) - B\mathrm{d}y)^2 + (A + B)\mathrm{d}y^2 + h_0.$$

Proposition 4.1.12. Let $(M, g) = (I, \varepsilon\mathrm{d}t^2) \times_a^b (N, h)$ be a tilted product, $\phi \in \mathfrak{F}(N)$ and $\tilde{g} = \mathrm{e}^{2\pi_2^*\phi}g$. Then, $(M, \tilde{g})$ is isometric to a tilted product.

Proof. Consider the transformation $t'(t, x) = t\mathrm{e}^{\phi(x)}$:

$$\tilde{g} = \varepsilon\mathrm{e}^{2\phi}(\mathrm{d}t - b)^2 + a^2\mathrm{e}^{2\phi}h = \varepsilon(\mathrm{d}t' - \tilde{b})^2 + a^2\tilde{h},$$

with $\tilde{b} = \mathrm{e}^\phi b - t'\mathrm{d}\phi$ and $\tilde{h} = \mathrm{e}^{2\phi}h$. □

Proposition 4.1.13. For $f \in \mathfrak{F}(N)$, the tilted product manifold $(M, g) = (I, \varepsilon\mathrm{d}t^2) \times_a^b (N, h)$ is isometric to $(M, g) = (I', \varepsilon\mathrm{d}t'^2) \times_a^{b+\mathrm{d}f} (N, h)$.

Proof. This is just a change of slicing, with $t'(t, x) = t + f(x)$: $\mathrm{d}t' - \mathrm{d}f = \mathrm{d}t$. $\square$

Corollary 4.1.14. *A tilted product manifold is isometric to a twisted product if the shift form is an exact form defined on the fiber. A shear-free kinematical manifold is locally isometric to a twisted product if the vorticity and the acceleration vanish.*

4.2 Causal Structure

The results in this section are expanded on in Section 6.1, where we exploit the gauge freedom of changing the slicing.

Proposition 4.2.1. *Let $(M, g) = (I, -\mathrm{d}t^2) \times_a^b (N, h)$ be a tilted twisted product spacetime and $p \in M$. Then, the hyperplane tangent to the slice N_s through p is*

- *timelike if and only if $|b_p| > a(p)$,*
- *null if and only if $|b_p| = a(p)$,*
- *spacelike if and only if $|b_p| < a(p)$.*

Proof. We notationally omit the point p. Let $\bar{b}$ be the unique vector tangent to N_s with $b = h(\bar{b}, \cdot)$. For any vector w tangent to N_s, we have

$$g(w, w) = -b(w)b(w) + a^2 h(w, w) = -h(\bar{b}, w)h(\bar{b}, w) + a^2 h(w, w)$$
$$\geq -h(\bar{b}, \bar{b})h(w, w) + a^2 h(w, w) = (a^2 - |b|^2)h(w, w),$$

where we have used the Cauchy–Schwarz inequality. The equality is attained for $w = \bar{b}$ and the result follows. $\square$

Proposition 4.2.2. *A tilted twisted product spacetime $(I, -\mathrm{d}t^2) \times_a^b (N, h)$ is stably causal if $|b| < a$. Suppose N is compact. Then, (M, g) is non-chronological (respectively, non-stably causal) if there exists a slice N_s with $|b|_p > a(p)$ (respectively, $|b|_p = a(p)$) for all $p \in N_s$.*

Proof. If $|b| < a$, then t is a temporal function:

$$g(\nabla t, \nabla t) = g^{-1}(\mathrm{d}t, \mathrm{d}t) = \frac{|b|^2 - a^2}{a^2} < 0.$$

If $|b| > a$ on a slice, this slice is a compact timelike hypersurface which must contain a closed timelike curve (Proposition 2.3.17).

Now suppose $|b| = a$ on a slice N_s. Consider the metric pertubation g_ϵ with shift form $b_\epsilon = (1 + \epsilon)b$, where $\epsilon \in \mathfrak{F}(I)$, $\epsilon \geq 0$ and $\epsilon(s) > 0$. We have that (M, g_ϵ) is non-causal, and $g - g_\epsilon = \epsilon(\mathrm{d}t \vee b - (2 + \epsilon)b \otimes b)$. In adapted coordinates, at some point (t, x),

$$\max_{ij}(|(g - g_\epsilon)_{ij}|) = \max_{ij} \epsilon(t)|t_i b_j + t_j b_i - (2 + \epsilon(t))b_i b_j|$$
$$\leq \max_{ij} \epsilon(t)(2|b_j| + (2 + \epsilon(t))|b_j|^2) \leq C(t, x)\epsilon(t)(2 + \epsilon(t))$$

for some function $C \colon M \to \mathbb{R}$, $C > 0$. Let δ be a continuous positive function on M. Since N is compact, the number $\delta^*(t) = \min_{x \in N_t} \delta(t, x)(C(t, x))^{-1}$ is strictly positive for any $t \in I$. If we choose ϵ so that $\epsilon(t)(2 + \epsilon(t)) < \delta^*(t)$ for every $t \in I$, the non-causal metric g_ϵ is contained in the δ-neighborhood of g. Thus, (M, g) fails to be stably causal. $\qquad\square$

Example 4.2.3. Given $\omega_0 > 0$, the non-chronological (in fact, totally vicious) 3-dimensional Gödel spacetime [Göd49] with metric

$$g = -\mathrm{d}t^2 - \mathrm{e}^{\sqrt{2}\omega_0 x}\mathrm{d}t \vee \mathrm{d}y + \mathrm{d}x^2 - \frac{1}{2}\mathrm{e}^{2\sqrt{2}\omega_0 x}\mathrm{d}y^2$$
$$= -(\mathrm{d}t + \mathrm{e}^{\sqrt{2}\omega_0 x}\mathrm{d}y)^2 + \mathrm{d}x^2 + \frac{1}{2}\mathrm{e}^{2\sqrt{2}\omega_0 x}\mathrm{d}y^2$$

defined on $M = \mathbb{R} \times \mathbb{R}^2$ is a tilted product $(M, g) = (\mathbb{R}, -\mathrm{d}t^2) \times^b_a (\mathbb{R}^2, h)$ with $h = \mathrm{d}x^2 + \frac{1}{2}\mathrm{e}^{2\sqrt{2}\omega_0 x}\mathrm{d}y^2$, $b = -\mathrm{e}^{\sqrt{2}\omega_0 x}\mathrm{d}y$ and $a = 1$. It holds that $|b| = \sqrt{2} > 1 = a$. Note that (N, h) is simply a hyperbolic plane with Gaussian curvature $\frac{\kappa^h}{2} = -2\omega_0^2$.

Moreover, employing the conformal change $g \mapsto \tilde{g} = 2\mathrm{e}^{-2\sqrt{2}\omega_0 x}g$, we arrive at the even simpler tilted product

$$\tilde{g} = -(\mathrm{d}y + 2\mathrm{e}^{-\sqrt{2}\omega_0 x}\mathrm{d}t)^2 + 2\mathrm{e}^{-2\sqrt{2}\omega_0 x}(\mathrm{d}x^2 + \mathrm{d}t^2),$$

where the coordinate function y also fails to be a temporal function, as expected.

We now set $\omega_0 = \sqrt{2}$ for convenience. Through the well-known coordinate transform $(t, x, y) \mapsto (t', r, \phi)$, defined on a dense subset $M_1 \subset M$, given by

$$\mathrm{e}^{2x} = \cosh(2r) + \cos\phi\sinh(2r),$$
$$\sqrt{2}y\mathrm{e}^{2x} = \sin\phi\sinh(2r),$$
$$\tan\left(\tfrac{1}{2}(\phi + \sqrt{2}(t - t'))\right) = \mathrm{e}^{-2r}\tan(\tfrac{1}{2}\phi),$$

we deduce yet another tilted product form of the Gödel metric:

$$g|_{M_1} = -(\mathrm{d}t' - \sqrt{2}\sinh^2 r\,\mathrm{d}\phi)^2 + \mathrm{d}r^2 + \sinh^2 r\cosh^2 r\,\mathrm{d}\phi^2.$$

In fact, this transformation is just a change of slicing, $t'(t, x) = t + f(x)$, or equivalently, a change of the shift form, $b' = b + \mathrm{d}f$ with $b' = \sqrt{2}\sinh^2 r \cdot \mathrm{d}\phi$. The

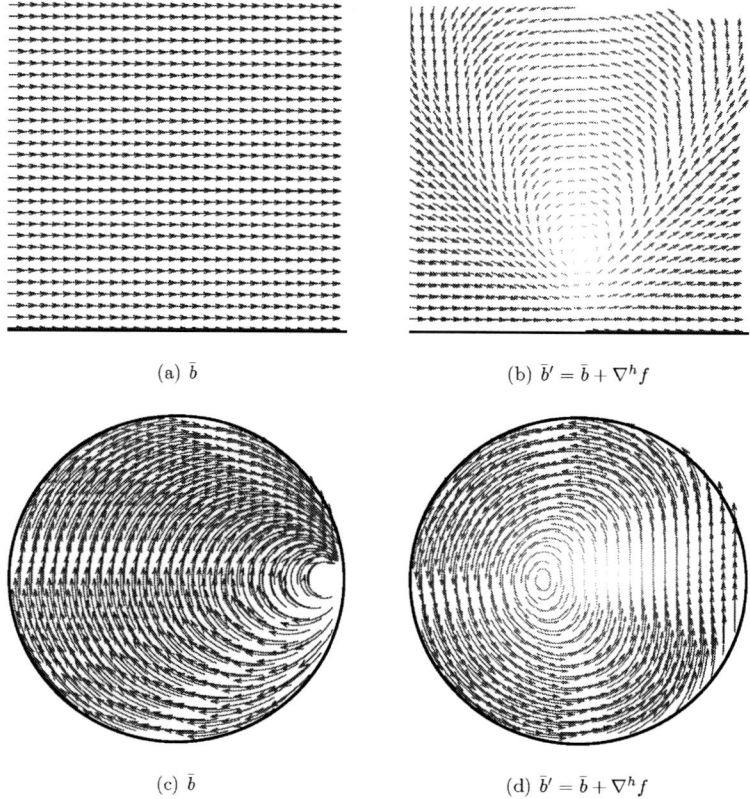

(a) $\bar{b}$

(b) $\bar{b}' = \bar{b} + \nabla^h f$

(c) $\bar{b}$

(d) $\bar{b}' = \bar{b} + \nabla^h f$

Figure 4.1: The shift vector field of the three-dimensional Gödel spacetime in two-dimensional hyperbolic space with two different choices of slicing, visualized in the Poincaré upper half-plane and the Poincaré disk.

spatial metric is the same as above, written in different coordinates. It becomes manifest that (M, g) is non-chronological: The integral curves of the lift of the vector field $\bar{b}'$ with $b' = h(\bar{b}', \cdot)$ are closed and timelike in the region with $|b'| > 1$, or $r > \operatorname{arctanh}(\frac{1}{\sqrt{2}}) = \log(1 + \sqrt{2})$. We may visualize the vector fields $\bar{b}'$ and $\bar{b}$ by transforming to well-known standard representations of the hyperbolic plane such as the Poincaré upper half-plane $H \subset \mathbb{C}$ or the Poincaré disk $D \subset \mathbb{C}$, see Figure 4.1. Note that the two representations are related via the Cayley transform $H \to D$, $z \mapsto \frac{z-i}{z+i}$. In particular the Poincaré disk visualization reveals that the integral curves of $\bar{b}$ already "close at infinity".

4.3 Ricci and Scalar Curvature

The $(D + 1)$-composition of the Ricci and scalar curvature of a tilted *warped* product (i.e., a tilted product metric (4.4) with a twisting function that is constant on the slices) is given by the following:

Proposition 4.3.1. *Consider a tilted warped product* $(M, g) = (I, \varepsilon dt^2) \times_a^b (N, h)$ *and introduce adapted coordinates* (t, x). *Then, the components of the Ricci tensor of* (M, g) *are given by:*

$$
\begin{aligned}
R_{ij} = \, & r_{ij} - \dot{b}_i \dot{b}_j - D_{(i}\dot{b}_{j)} - \ddot{b}_{(i}b_{j)} - (d-3)\theta D_{(i}b_{j)} - (d-5)\theta \dot{b}_{(i}b_{j)} \\
& + \frac{2\varepsilon}{a^2}\left[\bar{\omega}^k{}_j \omega_{ik} + (2\dot{\bar{b}}^k + (d-3)\theta \bar{b}^k)\omega_{k(j}u_{i)} + \bar{b}^k \dot{\omega}_{k(i}u_{j)} + u_{(j}J_{i)}\right] \\
& + \left[\frac{|\omega|^2}{a^4} - \frac{\varepsilon}{2a^2}\left((d-3)\theta(|b|^2)^{\bullet} + (|b|^2)^{\bullet\bullet} + 2D_k\dot{\bar{b}}^k\right) - (d-1)(\theta^2 + \dot{\theta})\right]u_i u_j \\
& + (d-3)(\theta^2 - \dot{\theta})b_i b_j - 2\varepsilon(d-2)\dot{\theta}u_{(i}b_{j)} + 2\varepsilon\theta u_{(i}\dot{b}_{j)} \\
& + \left[-\theta^2((d-3)|b|^2 + \varepsilon(d-1)a^2) - \dot{\theta}(|b|^2 + \varepsilon a^2) - \theta((|b|^2)^{\bullet} + D_k\bar{b}^k)\right]h_{ij}.
\end{aligned}
$$

The dots denote partial derivatives with respect to t, r_{ij} *are the components of the Ricci tensor of* (N, h), *and* D *is the Levi-Civita derivative of* h. *(Both lifted, i.e.,* $r_{i0} = r_{0i} = 0$ *and* $D_0 b_i = 0$ *is understood.) Furthermore,* $\theta := (\log a)^{\bullet}$ *is the Hubble-Lemaître parameter,* $\omega_{ij} = -\varepsilon(D_{[i}b_{i]} + \dot{b}_{[i}b_{j]})$ *the vorticity tensor[1] and* $J_i := D_k\bar{\omega}^k{}_i$. *The scalar curvature* R *of* (M, g) *computes to:*

$$
\begin{aligned}
a^2 R = \, & r - \varepsilon a^{-2}|\omega|^2 - (|b|^2)^{\bullet\bullet} - (2d-5)\theta(|b|^2)^{\bullet} - 2D_k\dot{\bar{b}}^k - 2(d-2)\theta D_k\bar{b}^k \\
& - \varepsilon(d-1)a^2(d\theta^2 + 2\dot{\theta}) - (d-2)|b|^2((d-3)\theta^2 + 2\dot{\theta}),
\end{aligned}
$$

where r *is the scalar curvature of* (N, h).

---

[1] Please note that the formula for the vorticity given in [GPS11], "$\omega_{ij} = -\varepsilon D_{[j}b_{i]}$", is false. Fortunately, the error is of no consequence for the rest of the paper.

Proof. If not otherwise noted, indices are raised and lowered via the metric g, partial derivatives are denoted by commas. Note that the Christoffel symbols of the Levi-Civita connection induced by the non-tilted, non-warped metric $\bar{g} = \varepsilon dt^2 + h$ may be written as

$$\bar{\Gamma}^k_{ij} = \frac{1}{2}\bar{h}^{kn}\left(h_{nj,i} + h_{ni,j} - h_{ij,n}\right).$$

Writing $f_{ij} := u_{i,j} - u_{j,i}$, the Christoffel symbols corresponding to g on the other hand are:

$$\begin{aligned}
\Gamma^k_{ij} &= \frac{1}{2}g^{kn}\left(g_{nj,i} + g_{ni,j} - g_{ij,n}\right) \\
&= \frac{1}{2}g^{kn}\left((\varepsilon u_n u_j + a^2 h_{nj})_{,i} + (\varepsilon u_n u_i + a^2 h_{ni})_{,j} - (\varepsilon u_i u_j + a^2 h_{ij})_{,n}\right) \\
&= \frac{\varepsilon}{2}\left(u^k(u_{j,i} + u_{i,j}) + u_j f^k_{\ i} + u_i f^k_{\ j}\right) \\
&\quad + \frac{1}{2}a^2 g^{kn}(h_{nj,i} + h_{ni,j} - h_{ij,n}) + a g^{kn}(a_{,i}h_{nj} + a_{,j}h_{ni} - a_{,n}h_{ij}).
\end{aligned}$$

Using the equation (4.5) for the inverse metric and plugging

$$\begin{aligned}
&\frac{1}{2}a^2 g^{kn}(h_{nj,i} + h_{ni,j} - h_{ij,n}) \\
&= \frac{1}{2}\bar{h}^{kn}(h_{nj,i} + h_{ni,j} - h_{ij,n}) - \frac{\varepsilon}{2}u^k(\bar{h}^{mn}u_m)(h_{nj,i} + h_{ni,j} - h_{ij,n}) \\
&= \bar{\Gamma}^k_{ij} - \varepsilon u^k u_n \bar{\Gamma}^n_{ij}
\end{aligned}$$

into this expression gives the following, denoting the Levi-Civita derivative with respect to $\bar{g}$ by a vertical bar:

$$\Gamma^k_{ij} = \bar{\Gamma}^k_{ij} + \frac{\varepsilon}{2}\left((u_{i|j} + u_{j|i})u^k + u_i f^k_{\ j} + u_j f^k_{\ i}\right) + a g^{kn}(a_{,i}h_{nj} + a_{,j}h_{ni} - a_{,n}h_{ij}).$$

Note that $t_{k|i} = \delta^0_{\ k|i} = 0$ (thus $u_{i|j} = -\varepsilon b_{i|j}$) and $u^k_{\ |i} = 0$. Since we assume that a is constant on the slices, we have $a_{,k} = \dot{a}t_k$ and may write

$$\Gamma^k_{ij} = \bar{\Gamma}^k_{ij} + \beta^k_{\ ij} + (\log a)^\bullet \Omega^k_{\ ij} \tag{4.6}$$

where

$$\begin{aligned}
\beta^k_{\ ij} &:= \frac{\varepsilon}{2}\left((u_{i|j} + u_{j|i})u^k + u_i f^k_{\ j} + u_j f^k_{\ i}\right), \\
\Omega^k_{\ ij} &:= a^2 g^{kn}(t_i h_{nj} + t_j h_{ni} - t_n h_{ij}).
\end{aligned}$$

These tensors may be contracted as follows:

$$\begin{aligned}
2\varepsilon\beta^k_{\ kj} &= (u_{j|k} + u_{k|j})u^k + u_k f^k_{\ j} + u_j f^k_{\ k} \\
&= (u_{j|k} + u_{k|j})u^k + u^k(u_{k|j} - u_{j|k}) = 0 \tag{4.7}
\end{aligned}$$

and

$$\Omega^k{}_{kj} = a^2 g^{kn}(t_k h_{nj} + t_j h_{nk} - t_n h_{kj})$$
$$= a^2(t^n h_{nj} + t_j h^k{}_k - t^k h_{kj}) = t_j(\delta^k{}_k - \varepsilon u^k u_k) = (d-1)t_j. \qquad (4.8)$$

Now the general formula for the components of the Ricci tensor in a given coordinate basis is

$$R_{ij} = \Gamma^k_{ij,k} - \Gamma^k_{ki,j} + \Gamma^k_{km}\Gamma^m_{ij} - \Gamma^k_{im}\Gamma^m_{kj}.$$

If we plug in (4.6) and account for (4.7), (4.8) the result is:

$$R_{ij} = \bar{R}_{ij} + \beta^k{}_{ij|k} - \beta^k{}_{im}\beta^m{}_{kj} + ((\log a)^\bullet)^2 \left((d-1)t_k\Omega^k{}_{ij} - \Omega^k{}_{im}\Omega^m{}_{kj}\right)$$
$$+ (\log a)^{\bullet\bullet} \left(t_k\Omega^k{}_{ij} - (d-1)t_i t_j\right)$$
$$+ (\log a)^\bullet \left(\Omega^k{}_{ij|k} + (d-1)t_k\beta^k{}_{ij} - \beta^k{}_{im}\Omega^m{}_{kj} - \beta^m{}_{kj}\Omega^k{}_{im}\right). \qquad (4.9)$$

We compute the missing terms. First, denoting $\dot{u}_i := u^k\nabla_k u_i = -\varepsilon u^k b_{i|k} =: -\varepsilon \dot{b}_i$,

$$2\varepsilon\beta^k{}_{ij|k} = \dot{u}_{j|i} + \dot{u}_{i|j} + u_{i|k}f^k{}_j + u_i f^k{}_{j|k} + u_{j|k}f^k{}_i + u_j f^k{}_{i|k}$$
$$= \dot{u}_{j|i} + \dot{u}_{i|j} + \frac{1}{2}(u_{i|k} - u_{k|i} + u_{k|i} + u_{i|k})f^k{}_j + \frac{1}{2}(u_{j|k} - u_{k|j} + u_{k|j} + u_{j|k})f^k{}_i$$
$$+ u_i f^k{}_{j|k} + u_j f^k{}_{i|k}$$
$$= f_{ik}f^k{}_j + u_i f^k{}_{j|k} + u_j f^k{}_{i|k} + \dot{u}_{j|i} + \dot{u}_{i|j}$$
$$+ \frac{1}{2}\left((u_{i|k} + u_{k|i})f^k{}_j + (u_{j|k} + u_{k|j})f^k{}_i\right)$$

and

$$4\beta^k{}_{im}\beta^m{}_{jk} = \left[(u_{i|m} + u_{m|i})u^k + u_m f^k{}_i + u_i f^k{}_m\right] \cdot \left[(u_{j|k} + u_{k|j})u^m + u_j f^m{}_k + u_k f^m{}_j\right]$$
$$= \dot{u}_j\dot{u}_i + u_j\dot{u}^m(u_{i|m} + u_{m|i}) + \varepsilon f^m{}_j(u_{i|m} + u_{m|i}) + \varepsilon f^k{}_i(u_{j|k} + u_{k|j})$$
$$- u_j\dot{u}_k f^k{}_i + \dot{u}_i\dot{u}_j + \dot{u}^k(u_{j|k} + u_{k|j})u_i + f^k{}_m f^m{}_k u_i u_j - \dot{u}^m f_{mj}u_i$$
$$= 2\dot{u}_i\dot{u}_j - f^2 u_i u_j + u_j\dot{u}^m(u_{i|m} + u_{m|i}) - u_j\dot{u}^m(u_{m|i} - u_{i|m})$$
$$+ u_i\dot{u}^m(u_{j|m} + u_{m|j}) - u_i\dot{u}^m(u_{m|j} - u_{j|m})$$
$$+ \varepsilon f^m{}_j(u_{i|m} + u_{m|i}) + \varepsilon f^m{}_i(u_{j|m} + u_{m|j})$$
$$= 2\dot{u}_i\dot{u}_j - f^2 u_i u_j + 2\dot{u}^m(u_{i|m}u_j + u_{j|m}u_i)$$
$$+ \varepsilon f^m{}_j(u_{i|m} + u_{m|i}) + \varepsilon f^m{}_i(u_{j|m} + u_{m|j})$$

with $f^2 := f_{mn}f^{mn}$. Thus,

$$
\begin{aligned}
\beta^k{}_{ij|k} - \beta^k{}_{im}\beta^m{}_{jk} &= \frac{\varepsilon}{2}(f_{ik}f^k{}_j + u_if^k{}_{j|k} + u_jf^k{}_{i|k} + \dot{u}_{j|i} + \dot{u}_{i|j}) \\
&\quad - \frac{1}{2}\dot{u}_i\dot{u}_j + \frac{1}{4}f^2u_iu_j - \frac{1}{2}\dot{u}^m(u_{i|m}u_j + u_{j|m}u_i) \\
&= \frac{\varepsilon}{2}(f_{ik}f^k{}_j + u_if^k{}_{j|k} + u_jf^k{}_{i|k} + \dot{u}^m(b_{i|m}u_j + b_{j|m}u_i)) \\
&\quad - \frac{1}{2}(\dot{b}_{j|i} + \dot{b}_{i|j}) - \frac{1}{2}\dot{b}_i\dot{b}_j + \frac{1}{4}f^2u_iu_j.
\end{aligned}
\tag{4.10}
$$

Note that $2\omega_{ij} = a^4 h_i{}^k h_j{}^l f_{kl}$, and we have $f_{ij} = 2\omega_{ij} + u_i\dot{b}_j - u_j\dot{b}_i$ (in particular, $u^n f_{ni} = \varepsilon \dot{b}_i$) and

$$
\begin{aligned}
a^2 f^k{}_i &= (\bar{h}^{kn} + (|b|^2 + \varepsilon a^2)u^k u^n + u^k\bar{b}^n + u^n\bar{b}^k)f_{ni} \\
&= \bar{h}^{kn}(2\omega_{ni} + u_n\dot{b}_i - u_i\dot{b}_n) + \varepsilon(|b|^2 + \varepsilon a^2)u^k\dot{b}_i + u^k\bar{b}^n(2\omega_{ni} + u_n\dot{b}_i - u_i\dot{b}_n) + \varepsilon\bar{b}^k\dot{b}_i \\
&= 2\bar{h}^{kn}\omega_{ni} + 2u^k\bar{b}^n\omega_{ni} + a^2 u^k\dot{b}_i - \bar{b}^n\dot{b}_n u^k u_i - \dot{\bar{b}}^k u_i.
\end{aligned}
$$

Via similar calculations, this implies with $|\omega|^2 = \bar{h}^{im}\bar{h}^{jn}\omega_{ij}\omega_{mn}$:

$$
a^2 f_{ik}f^k{}_j = 4\bar{h}^{kn}\omega_{nj}\omega_{ik} + 2\dot{\bar{b}}^n(\omega_{nj}u_i + \omega_{ni}u_j) - |\dot{b}|^2 u_i u_j - \varepsilon a^2 \dot{b}_i\dot{b}_j
\tag{4.11}
$$

and

$$
f^2 = \frac{4}{a^4}|\omega|^2 + \frac{2\varepsilon}{a^2}|\dot{b}|^2.
\tag{4.12}
$$

Furthermore note with

$$
t_k f^k{}_i = \frac{1}{a^2}(2\bar{b}^n\omega_{ni} - (\bar{b}^n\dot{b}_n)u_i) + \dot{b}_i,
\tag{4.13}
$$

we have

$$
\begin{aligned}
a^2 f^k{}_{i|k} &= -2a^2(\log a)^\bullet t_k f^k{}_i + \left[2\bar{h}^{kn}\omega_{ni} + 2u^k\bar{b}^n\omega_{ni} + a^2 u^k\dot{b}_i - \bar{b}^n\dot{b}_n u^k u_i - \dot{\bar{b}}^k u_i\right]_{|k} \\
&= -4(\log a)^\bullet\bar{b}^n\omega_{ni} + 2(\log a)^\bullet(\bar{b}^n\dot{b}_n)u_i + 2\bar{h}^{kn}\omega_{ni|k} + 2\bar{b}^n\dot{\omega}_{ni} + 2\dot{\bar{b}}^n\omega_{ni} \\
&\quad + a^2\ddot{b}_i - (\bar{b}^n\dot{b}_n)^\bullet u_i + \varepsilon(\bar{b}^n\dot{b}_n)\dot{b}_i - (\dot{\bar{b}}^k{}_{|k})u_i + \varepsilon\dot{\bar{b}}^k b_{i|k}.
\end{aligned}
\tag{4.14}
$$

Also,

$$
\begin{aligned}
a^2 \dot{u}^k(b_{i|k}u_j + b_{j|k}u_i) &= -\varepsilon(\bar{h}^{kn} + (|b|^2 + \varepsilon a^2)u^k u^n + u^k\bar{b}^n + u^n\bar{b}^k)(b_{i|k}u_j + b_{j|k}u_i)\dot{b}_n \\
&= -\varepsilon(\dot{\bar{b}}^k + (\bar{b}^n\dot{b}_n)u^k)(b_{i|k}u_j + b_{j|k}u_i) \\
&= -\varepsilon\dot{\bar{b}}^k(b_{i|k}u_j + b_{j|k}u_i) - \varepsilon(\bar{b}^n\dot{b}_n)(\dot{b}_i u_j + \dot{b}_j u_i).
\end{aligned}
\tag{4.15}
$$

Plugging equations (4.11)–(4.15) into (4.10), we get with $\bar{\omega}^i{}_j := \bar{h}^{in}\omega_{nj}$:

$$\beta^k{}_{ij|k} - \beta^k{}_{im}\beta^m{}_{jk} = -\dot{b}_i\dot{b}_j - \frac{1}{2}(\dot{b}_{i|j} + \dot{b}_{j|i}) + \frac{\varepsilon}{2}(\ddot{b}_i u_j + \ddot{b}_j u_i) + \frac{|\omega|^2}{a^4}u_i u_j$$

$$+ \frac{\varepsilon}{a^2}\Big[2\bar{\omega}^k{}_j\omega_{ik} + (2(\log a)^\bullet \bar{b}^n \dot{b}_n - (\bar{b}^n \dot{b}_n)^\bullet - \dot{\bar{b}}^k{}_{|k})u_i u_j$$

$$+ 2(\dot{\bar{b}}^n - (\log a)^\bullet \bar{b}^n)(\omega_{nj}u_i + \omega_{ni}u_j) + \bar{b}^n(\dot{\omega}_{ni}u_j + \dot{\omega}_{nj}u_i)$$

$$+ \bar{\omega}^k{}_{i|k}u_j + \bar{\omega}^k{}_{j|k}u_i\Big] \tag{4.16}$$

The following formulas may be easily found by realizing that $b_i = a^2 t_k h^k{}_i = a^2 b_k h^k{}_i$:

$$(d-1)t_k\Omega^k{}_{ij} - \Omega^k{}_{im}\Omega^m{}_{jk} = (d-3)b_i b_j - (d-1)u_i u_j$$
$$- (d-3)(|b|^2 + \varepsilon a^2)h_{ij}, \tag{4.17}$$

$$t_k\Omega^k{}_{ij} - (d-1)t_i t_j = -(d-3)b_i b_j - (d-1)u_i u_j - (|b|^2 + \varepsilon a^2)h_{ij}$$
$$- \varepsilon(d-2)(u_i b_j + b_i u_j). \tag{4.18}$$

As for the remaining terms, those compute to:

$$\Omega^k{}_{ij|k} = \dot{b}_i b_j + b_j \dot{b}_i + \varepsilon(u_i \dot{b}_j + \dot{b}_i u_j) - \Big(2\bar{b}^n \dot{b}_n + 2\varepsilon a\dot{a} + \bar{b}^k{}_{|k}\Big)h_{ij}, \tag{4.19}$$

$$t_m\beta^m{}_{ij} = -\frac{1}{2}(b_{i|j} + b_{j|i}) + \frac{\varepsilon}{a^2}\bar{b}^n(\omega_{ni}u_j + \omega_{nj}u_i)$$

$$+ \frac{\varepsilon}{2}(\dot{b}_i u_j + u_i \dot{b}_j) - \frac{\varepsilon}{a^2}(\bar{b}^n \dot{b}_n)u_i u_j, \tag{4.20}$$

$$\beta^m{}_{ik}\Omega^k{}_{mj} = -b_{j|i}. \tag{4.21}$$

We note that $\bar{R}_{ij} = r_{ij}$ and $b_{i|j} = D_j b_i + \dot{b}_i t_j$. With this, plugging (4.16)–(4.21) into (4.9) finally yields the expression of the Ricci tensor; the scalar curvature of course follows from computing the trace (with respect to g). $\qquad\square$

Remark 4.3.2. Note that while the above formulas do not apply to the more complicated general shear-free case $a = a(t, x)$, by setting $\theta = 0$ they describe the curvature for any rigidly rotating reference frame.

Corollary 4.3.3. *Suppose the Einstein–Hilbert equation $R_{ij} - \frac{R}{2}g_{ij} = T_{ij}$ holds for a tilted warped product spacetime $(I, -dt^2) \times^b_a (N, h)$ with $d = 4$. Let the usual $(3+1)$-decomposition of the energy–momentum tensor be given by*

$$T_{ij} = \rho u_i u_j + pP_{ij} + 2q_{(i}u_{j)} + \Pi_{ij},$$

with energy density ρ, (effective) pressure p, heat flux q_i and anisotropic pressure Π_{ij}.

Then,

$$a^2\rho = \frac{r}{2} + \frac{3|\omega|^2}{2a^2} - \theta(|b|^2)^\bullet - 2\mathrm{D}_k\bar{b}^k + 3a^2\theta^2 - |b|^2(\theta^2 + 2\dot{\theta}),$$

$$3a^2p = -\frac{r}{2} + \frac{|\omega|^2}{2a^2} + 2\theta(|b|^2)^\bullet + (|b|^2)^{\bullet\bullet} + 2\mathrm{D}_k\dot{\bar{b}}^k$$
$$+ 2\theta\mathrm{D}_k\bar{b}^k - 9a^2\theta^2 - 6a^2\dot{\theta} + |b|^2(\theta^2 + 2\dot{\theta}),$$

$$q_i = -\frac{2}{a^2}\left((2\dot{\bar{b}}^k + \theta\bar{b}^k)\omega_{ki} + \bar{b}^k\dot{\omega}_{ki} + J_i\right) + 4\dot{\theta}b_i - 2\theta\dot{b}_i,$$

$$\Pi_{ij} = r_{ij} - \dot{b}_i\dot{b}_j - \mathrm{D}_{(i}\dot{b}_{j)} - \ddot{b}_{(i}b_{j)} - \theta\mathrm{D}_{(i}b_{j)} + \theta\dot{b}_{(i}b_{j)}$$
$$- \frac{2}{a^2}\bar{\omega}^k{}_j\omega_{ik} + (\theta^2 - \dot{\theta})b_ib_j - \frac{\alpha}{3}h_{ij},$$

where the function α is determined by the condition $\bar{h}^{ij}\Pi_{ij} = 0$.

Proof. This is a straightforward computation using the formulas

$$\rho = R_{kl}u^k u^l + \frac{R}{2}, \quad 3p = R_{kl}u^k u^l - \frac{R}{2},$$
$$q_i = a^2 h^k{}_i R_{kl}u^l, \quad \Pi_{ij} = a^4 h^k{}_i h^l{}_j R_{kl} - (\dots)h_{ij}.$$

These are essentially coordinate expressions for Eqs. (3.2)–(3.5) for $d = 4$. $\qquad\Box$

4.4 Symmetries

For tilted warped products, the condition that the standard observer field is conformally stationary can be expressed in terms of the shift form and the scale parameter:

Proposition 4.4.1. *The standard reference frame $V = \partial_t$ of a tilted warped product $(I, \varepsilon\mathrm{d}t^2) \times^b_a (N, h)$ is (locally) conformally stationary if and only if the shift is of the form*

$$b = a \cdot \left(\xi + \int_0^t \frac{\mathrm{d}\tau}{a(\tau)} \cdot \chi\right),$$

where $\chi \in \Gamma(TN)$ is (locally) exact, and $\xi \in \Gamma(TN)$.

Proof. From Proposition 3.3.16 we know that a shear-free reference frame is parallel to a (local) conformal vector field if and only if its red-shift one-form is (locally) exact. For tilted warped products, we have $\ell_i = -\varepsilon\dot{u}_i + \varepsilon\theta u_i = \dot{b}_i + \theta(t_i - b_i)$, and this condition translates to

$$0 = \ell_{[i|j]} = \dot{b}_{[i|j]} + \dot{\theta}t_{[i}b_{j]} - \theta b_{[i|j]}. \tag{4.22}$$

Transvecting this equation with u^i implies

$$\ddot{b}_j = (\theta b_j)^\bullet. \tag{4.23}$$

Substituting this equation and $b_{i|j} = D_j b_i + \dot{b}_i t_j$ into Eq. (4.22) yields

$$D_{[i}\dot{b}_{j]} = \theta D_{[i}b_{j]}. \tag{4.24}$$

Conversely, it is also not difficult to check that Eqs. (4.23) and (4.24) imply Eq. (4.22). Now, Eq. (4.23) may be readily integrated to give the desired expression for b_j. Substituting this expression into Eq. 4.24 yields the additional constraint $D_{[i}\chi_{j]} = 0$. $\quad\square$

Thus, examples of tilted warped product spacetimes with conformally stationary reference frame may be easily constructed. Such examples are of physical importance since these models are parallax-free [HP88]. If $b = a\xi$ and the metric $\gamma = h + \xi \otimes \xi$ is a homogeneous Riemannian metric on the fiber, the resulting conformally stationary spacetime is a Korotky–Obukhov model [Obu00]. In the next chapter, we discuss certain models where $b = a\xi$, and h is the metric of a round sphere (and thus homogeneous). In these models, the quadratic form induced by the tensor $\gamma = h + \xi \otimes \xi$ on the other hand does not even have to be positive-definite.

Furthermore, a symmetry of the fiber (N, h) may be inherited by the total space via the following construction, an example of which is presented in the next chapter:

Proposition 4.4.2. *Let (N, h) be a connected Riemannian manifold, $\dim(N) \geq 1$, which admits a geodesic Killing vector field K. Furthermore, let $a\colon I \to \mathbb{R}$ be an arbitrary positive function. Then, $(M, g) = (I, \varepsilon dt^2) \times_a^b (N, h)$ where $b = a \cdot h(K, \cdot)$ has the following properties:*

1. *The standard reference frame $V = \partial_t$ is conformally stationary.*

2. *The vector field $\hat{K}$ is Killing on (M, g).*

Note that $|K| = |K|_h$ is a constant. In the Lorentzian case $\varepsilon = -1$, we also have the following:

3. *The Killing vector field $\hat{K}$ is timelike (spacelike, null) if $|K| > 1$ ($|K| < 1$, $|K| = 1$). If it is causal, it is past-pointing. If it is spacelike and non-zero, it is nowhere tangent to a slice.*

4. *If $|K| < 1$, (M, g) is stably causal. Suppose N is compact. If $|K| > 1$ holds, (M, g) is totally vicious. If $|K| = 1$ holds, (M, g) is non-stably causal.*

Proof. We have indeed that $|K|$ is a constant: for any $X \in \Gamma(TN)$,

$$X(h(K, K)) = 2h(D_X K, K) = -2h(D_K K, X) = 0.$$

1. This is a direct consequence of Proposition 4.4.1 (and is independent of the fact that K is Killing).

2. Since K is geodesic, we have $\bar{b}^j D_j b_i = 0$. Since K is also Killing, we have $D_i b_j + D_j b_i = 0$ and $2\bar{b}^n \omega_{ni} = -\varepsilon \bar{b}^n (D_n b_i - D_i b_n) = -2\varepsilon \bar{b}^n D_n b_i = 0$. With these conditions, and the equations established in the proof of Proposition 4.3.1 relating the connections D and ∇, we find

$$\nabla_i u_j + \nabla_j u_i = 2\dot{a}a^{-1}h_{ij} - (\dot{b}_i u_j + \dot{b}_j u_i),$$
$$\nabla_i b_j + \nabla_j b_i = \varepsilon(|b|^2)^\bullet a^{-2} u_i u_j + 2\dot{a}a^{-1}(h_{ij} - (t_i b_j + t_j b_i)) + \dot{b}_i t_j + \dot{b}_j t_i.$$

With

$$\hat{K}_i = g_{ij}\hat{K}^j = a^{-1}g_{ij}\bar{b}^j = -a^{-1}|b|^2 u_i + ab_i = a(-|K|^2 u_i + b_i)$$

we may then compute

$$\nabla_i \hat{K}_j + \nabla_j \hat{K}_i = \dot{a}(-|K|^2(u_i t_j + u_j t_i) + (b_i t_j + b_j t_i))$$
$$- |K|^2 a(\nabla_i u_j + \nabla_j u_i) + a(\nabla_i b_j + \nabla_j b_i),$$

which, also noting that $\dot{b} = \dot{a}a^{-1}b$, yields the claim $\nabla_i \hat{K}_j + \nabla_j \hat{K}_i = 0$ after a short computation.

3. It is easy to check that $g(\hat{K}, \hat{K}) = a^2|K|^2(1 + \varepsilon|K|^2)$ and $g(\hat{K}, V) = -\varepsilon a|K|^2$ hold.

4. This is a direct consequence of Proposition 4.2.2 and Proposition 3.3.22.

$\square$

Chapter 5

EXAMPLES OF TILTED WARPED PRODUCT SPACETIMES

The standard cosmological model assumes—as a good approximation in later epochs, at least—that the large scale structure of the observable universe can be described by a perfect fluid solution to the Einstein–Hilbert equation in form of a Friedmann–Lemaître–Robertson–Walker (FLRW) metric, i.e., a warped product $(I, -\mathrm{d}t^2) \times_a (N, h)$, where (N, h) is a space of constant curvature. If the curvature of (N, h) is positive (negative, zero), one usually refers to the closed (open, flat) FLRW model [Lem31, Rob33, Wal37, Fri99].

With the groundwork laid so far in this thesis, a natural generalization of the FLRW metric would be to tilt it by introducing a non-zero shift form. In general, this construction does not result in a perfect fluid solution. However, it turns out that the closed FLRW model may be tilted to yield the following generalization of a perfect fluid:

Definition 5.0.3. Formally, a cosmological model (M, g, V) is a *viscous fluid* if

$$\mathrm{Ric} - \frac{\kappa}{2}g = \rho V^\flat \otimes V^\flat + p^* P + q \vee V^\flat - 2\eta\sigma \tag{5.1}$$

holds where $\rho, p^*, \eta \in \mathfrak{F}(M)$ are the *energy density*, *effective pressure* and *shear viscosity coefficient*, respectively ($\eta \geq 0$); $q \in \Gamma(HM^*)$ is the *heat flux*.

The fluid is called *perfect* if $q = 0$ and $\eta\sigma = 0$ hold. A perfect fluid with $p^* = 0$ is called *dust*.

Remark 5.0.4. For an overview of relativistic fluid dynamics we refer to [AC07] and the references therein.

5.1 The Tilted Closed FLRW Model: Conformally Stationary Case

On (a dense subset of) the unit 3-sphere $S^3 \subset \mathbb{R}^4$ set up coordinates (r, μ, ν) defining the parametrization [Wil70]:

$$[0,1] \times [0, 2\pi]^2 \to S^3, \ (r, \mu, \nu) \mapsto (r \cos \mu, r \sin \mu, \sqrt{1 - r^2} \cos \nu, \sqrt{1 - r^2} \sin \nu).$$

Now consider the manifold $M = I \times S^3$ with a suitable open interval $I \subset \mathbb{R}$, equipped with the tilted warped product metric

$$g = -(\mathrm{d}t - b_0 a(r^2 \mathrm{d}\mu + (1 - r^2)\mathrm{d}\nu))^2 + a^2(\frac{1}{1 - r^2}\mathrm{d}r^2 + r^2 \mathrm{d}\mu^2 + (1 - r^2)\mathrm{d}\nu^2), \quad (5.2)$$

where $b_0 \in \mathbb{R}$ and $a > 0$ is a function defined on I. The spatial metric h is just the canonical round metric on S^3. Provided that $b_0 \neq 0$, the shift one-form b does not vanish anywhere and is in fact, for each fixed $s \in I$, h-metrically equivalent to a Hopf vector field on S^3, $\bar{b} = b_0 a(s)(\partial_\mu + \partial_\nu)$. The spacetime (M, g) generalizes the closed FLRW spacetime (assuming a simply connected, complete fiber). It is a Korotky–Obukhov model of Bianchi type IX – for another model of this type which is a solution to the vacuum Einstein–Hilbert equation with cosmological constant, see [COS02].

Proposition 5.1.1. *The conformally stationary tilted closed FLRW spacetime* $(I \times S^3, g)$ *is stably causal for* $|b_0| < 1$ *and totally vicious for* $|b_0| > 1$. *It is non-stably causal for* $|b_0| = 1$.

Proof. Apply Proposition 4.4.2 to $K = b_0(\partial_\mu + \partial_\nu)$. □

Remark 5.1.2. For $|b_0| = 1$, the vector field $K = b_0(\partial_\mu + \partial_\nu)$ is null and Killing, according to Proposition 4.4.2. In fact, K is also parallel which means that (M, g) is a Brinkmann space [Bri25, BSS11]. Brinkmann spaces are also called pp-waves by some authors [KSMH79], while others refer to pp-waves as special Brinkmann spaces (see, for example, [Lei06]).

Proposition 5.1.3. *The conformally stationary tilted closed FLRW spacetime is a viscous fluid with respect to the standard observer field* $V = \partial_t$ *if and only if* $b_0 = 0$ *or the scale parameter is given by either*

$$a_-(t) = \alpha \sinh\left(\frac{t - t_0}{\alpha}\right), \qquad\qquad t_0 < t,$$

$$a_0(t) = t - t_0, \qquad\qquad t_0 < t,$$

$$a_+(t) = \frac{1}{\omega}\sin(\omega(t - t_0)), \qquad\qquad t_0 < t < t_0 + \frac{\pi}{\omega},$$

where t_0, α, ω *are constants of integration.*

Proof. Note that the standard FLRW model (i.e., the case $b_0 = 0$) is a perfect fluid for any scale parameter. One may also set $b_i = 0$ in the formulas provided by Corollary 4.3.3 to see this. Now, let $b_0 \neq 0$. Since the model is shear-free, the form of the energy–momentum tensor (5.1) enforces that the anisotropic pressure vanishes. The physical idea behind this condition is that the fluid shows resistance to shear stress via the friction of neighboring fluid layers producing viscous forces, which introduce a traceless tensor part to energy–momentum, i.e., anisotropic pressure. Since h is an Einstein metric (i.e., $r_{ij} = \frac{r}{3}h_{ij}$), and $\bar{b}$ is a geodesic Killing vector field on the slices (i.e., $D_{(i}b_{j)} = 0$), it may be readily verified via Corollary 4.3.3 that this condition is equivalent to

$$b_i b_j(-\ddot{a}a + (\dot{a})^2) - \bar{\omega}^k{}_j \omega_{ik} = \frac{1}{3}(|b|^2(-\ddot{a}a + (\dot{a})^2) + |\omega|^2)h_{ij}.$$

Noting that $\omega_{ij} = D_{[j}b_{i]}$, this equation may be evaluated in a straightforward manner. One then sees that the anisotropic pressure vanishes if and only if the scale parameter satisfies the differential equation

$$\ddot{a}a - (\dot{a})^2 + 1 = 0. \tag{5.3}$$

This equation may be solved analytically via the substitution $a = e^f$, yielding the autonomous equation $\ddot{f} + e^{-2f} = 0$ the integration of which is elementary. $\square$

Remark 5.1.4. In any case, the spacetime features a singularity. These three classes of solutions may be characterized by the deceleration parameter, $-\frac{a\ddot{a}}{\dot{a}^2}$, which is negative (zero, positive) for a_- (a_0, a_+). Taking limits of the solution parameters/constants of integration, we see that $\lim_{\alpha \to \infty} a_-(t) = \lim_{\omega \to 0} a_+(t) = a_0(t)$.

Introducing vorticity in the way done here thus yields a tight constraint that seriously limits cosmic evolution and matter models. We would also like to note that the model is purely Weyl electric [KSMH79, p. 51], i.e., $C_{iabc}\epsilon_j{}^{dbc}u^a u_d = 0$.

Definition 5.1.5. Let (M, g) be a spacetime.

1. The *dominant energy condition* is satisfied if $\text{Ric}(X, Y) \geq \frac{\kappa}{2}g(X, Y)$ holds for any future-pointing timelike vector fields X and Y.

2. The *weak energy condition* is satisfied if $\text{Ric}(X, X) \geq \frac{\kappa}{2}g(X, X)$ holds for any timelike vector field X.

3. The *strong energy condition* or *timelike convergence condition* is satisfied if for any timelike vector field X, $\text{Ric}(X, X) \geq 0$ holds.

Remark 5.1.6. See [Ren10, Section 4.1] for the above characterization of the dominant energy condition. Suppose the Einstein–Hilbert equation is satisfied for an energy–momentum tensor $T \in \Gamma(TM^* \vee TM^*)$. Physically, the energy conditions may be interpreted as follows (see, for example, [HE73, Section 4.3] for more details):

- **Dominant energy condition.** The dominant energy condition holds if and only if the energy flow vector field $T(X, \cdot)^\sharp$ is a future-pointing causal vector field whenever X is a future-pointing causal vector field. Therefore, no observer may measure an energy flux that locally propagates at a speed faster than the speed of light, or that transports matter/energy backward in time.

- **Weak energy condition.** For any observer in any reference frame W, the measured energy density $\rho_W = T(W, W)$ is non-negative.

- **Timelike convergence condition.** Cf. also [O'N83, p. 215 ff., p. 336 ff.]. Given a point $p \in M$ and an instantaneous observer $v \in T_p M$, denote the restspace of v by $\langle v \rangle^\perp$. Then, for any $x \in \langle v \rangle^\perp$, the vector $F_v(x) := \mathcal{R}_p(x, v)v \in \langle v \rangle^\perp$ determines the tidal acceleration of test particles moving in the direction of x. Therefore, the condition that $g_p(F_v(x), x) \leq 0$ holds at any point p and for all $v \in T_p M$ and $x \in \langle v \rangle^\perp$ may be interpreted as the physical statement that gravity always attracts. Since $\text{Trace}(F_v) = -\text{Ric}_p(v, v)$, the timelike convergence condition means that *on average*, gravity is an attractive force. Note that we assume that the cosmological constant is zero, or absorbed in the energy–momentum tensor; otherwise the strong energy condition and the timelike convergence condition are not equivalent.

Despite the name, the strong energy condition does not imply the weak energy condition. However, the weak energy condition is implied by the dominant energy condition. The null energy condition mentioned in Example 3.1.4 is implied by both the strong and the weak energy condition, respectively.

Proposition 5.1.7. *For the conformally stationary tilted closed FLRW model, the dominant and strong energy conditions are satisfied if $a = a_0$ or $a = a_+$. In the case $a = a_-$, the strong energy condition is satisfied if and only if $b_0 \geq 1$, and the dominant and weak energy conditions are satisfied if and only if $b_0 \leq 1$.*

Proof. This is a straightforward computation using Corollary 1 in [KST88]. □

Using Eq. (5.3) to eliminate second derivatives of a, the remaining components of the energy–momentum tensor are:

$$\rho = a^{-2} \left(-3(\dot{a})^2 (b_0^2 - 1) + 5b_0^2 + 3 \right), \tag{5.4}$$
$$p^* = a^{-2} \left(3(\dot{a})^2 (b_0^2 - 1) - b_0^2 + 1 \right), \tag{5.5}$$
$$q = -4a^{-2}b.$$

Note that $\rho + p^* = 4a^{-2}(b_0^2 + 1) \neq 0$ holds.

The following thermodynamical conditions may be stipulated [Col89]:

Definition 5.1.8. Formally, we say that a viscous fluid satisfies the *laws of linear thermodynamics* if

$$p^* = p - \zeta\Theta,$$

$$\operatorname{div}(n \cdot V) = 0, \qquad\qquad \text{particle number conservation law}, \quad (5.6)$$

$$\mathrm{d}\left(\frac{S}{n}\right) = \frac{1}{T}\mathrm{d}\left(\frac{\rho}{n}\right) + \frac{p}{T}\mathrm{d}\left(\frac{1}{n}\right), \qquad\qquad \text{Gibbs' relation}, \quad (5.7)$$

$$q = -kP\left(\nabla T + T\nabla_V V, \cdot\right), \qquad\qquad \text{temperature gradient law}, \quad (5.8)$$

where the functions $\zeta, n, k, T \geq 0$ are the *bulk viscosity coefficient, particle number density, thermal conductivity* and *temperature*, respectively. The functions p and S are the *pressure* and *entropy density*.

Remark 5.1.9. These conditions are designed so that the second law of thermodynamics, $\operatorname{div}(SV + \frac{1}{T}q^\sharp) \geq 0$, holds. Note, however, that more recent treatments of relativistic thermodynamics suggest that these constraints are too simple to give a full account of the actual physics. For an overview of different approaches to the problem, cf. for example [AC07].

By using the ansatz of a vanishing bulk viscosity, an equation of state of the form $S = \frac{\rho+p}{T}$, and matter variables that are constant on the slices, we derive the following model:

Proposition 5.1.10. *Assuming $a = a_0$ or $a = a_-$, the conformally stationary tilted closed FLRW model satisfies the laws of linear thermodynamics with*

$$\zeta = 0, \quad n = n_0 a^{-3}, \quad T = T_0 \cdot a^{-\frac{b_0^2-1}{b_0^2+1}},$$

$$S = \frac{4}{T_0}(b_0^2 + 1)a^{-\frac{b_0^2+3}{b_0^2+1}}, \quad k = \frac{2}{T_0 \dot{a}}(b_0^2 + 1)a^{-\frac{2}{b_0^2+1}},$$

where $n_0, T_0 \geq 0$ are constants. In this case, the equation of state $S(\rho, p, T) = \frac{\rho+p}{T}$ holds and entropy production is given by

$$\operatorname{div}(SV + \frac{1}{T}q^\sharp) = \frac{8b_0^2}{T_0(b_0^2 + 1)}\dot{a}a^{-\frac{2(b_0^2+2)}{b_0^2+1}}.$$

Proof. Like in the non-tilted case, particle conservation (5.6) enforces $n \propto a^{-3}$. If $\zeta = 0$, the first equation of state follows immediately from Eqs. (5.4), (5.5). Accounting for (5.3), we may solve (5.8) for the thermal conductivity: $k = \frac{4}{a(aT)^\bullet}$. With the above constraints, Gibbs' relation (5.7) also simplifies considerably and may be solved with respect to the temperature. The remaining formulas also follow from straightforward computations which may be facilitated by a computer algebra system since the metric is given explicitly in coordinates. In order for the thermal conductivity to be well-defined and non-negative, the expansion must be positive: $\dot{a} > 0$. $\square$

By similar means, we may assume a vanishing pressure to write down viscous dust solutions:

Proposition 5.1.11. *Assuming* $a = a_0$ *or* $a = a_-$, *and* $b_0 \leq 1$, *the conformally stationary tilted closed FLRW model satisfies the laws of linear thermodynamics with*

$$p = 0, \quad n = n_0 a^{-3}, \quad T = T_0, \quad \zeta = \frac{1 - b_0^2}{3a\dot{a}}(3(\dot{a})^2 - 1)$$

$$S = \frac{1}{a^2 T_0}\left(3(1 - b_0^2)(\dot{a})^2 + 3b_0^2 + 5\right), \quad k = \frac{4}{T_0 a \dot{a}},$$

where $n_0, T_0 \geq 0$ *are constants. The equation of state* $S(\rho, p, T) = \frac{\rho + p}{T}$ *holds, and entropy production is given by*

$$\operatorname{div}(SV + \frac{1}{T}q^\sharp) = \frac{\dot{a}}{a^3 T_0}\left(9(1 - b_0^2)(\dot{a})^2 + b_0^2 + 3\right).$$

5.1.1   Geodesics and Observations.

We wish to show some physical features of the (expanding) tilted FLRW universe a fluid observer travelling along an integral curve of $V = \partial_t$ would experience. First, an orthonormal tetrad is given by

$$\eta_0 = \partial_t, \quad \eta_1 = b_0 \partial_t + a^{-1}(\partial_\mu + \partial_\nu),$$

$$\eta_2 = \frac{1}{ar\sqrt{1-r^2}}(-(1 - r^2)\partial_\mu + r^2\partial_\nu), \quad \eta_3 = \frac{\sqrt{1-r^2}}{a}\partial_r.$$

Note that η_1 in fact points in the direction of the fluid's acceleration, and along its rotational axis. The integral curves of the vector fields

$$\xi_\pm = a^{-1}(\eta_0 \pm \eta_1) = a^{-1}((1 \pm b_0)\partial_t \pm a^{-1}(\partial_\mu + \partial_\nu))$$

are future-directed null geodesics, i.e., worldlines of light rays. The integral curves of $\partial_\mu + \partial_\nu$ may be visualized as simple closed curves running around the 2-dimensional tori embedded in S^3, given by $r = \text{const.}$

Without loss of generality, we assume $t_0 = 0$ and $b_0 \geq 0$. For the case $a = a_0$, the integral curves of $\xi_\pm$ are given by

$$t(\tau) = \sqrt{2(1 \pm b_0)\tau + t(0)^2}, \quad r(\tau) = \text{const.},$$

$$\mu(\tau(t)) = \nu(\tau(t)) = \pm(1 \pm b_0)^{-1}\log\left(\frac{t(\tau)}{t(0)}\right),$$

provided that $b_0 \neq 1$. Here, we assumed $\mu(0) = \nu(0) = 0$. If $b_0 = 1$, the null geodesics determined by ξ_- are closed for any scale parameter $a(t)$. This implies that a hypothetical human observer travelling along the fluid flow and constantly looking along the direction $\mp\eta_1$ would, at any time $t = t_2$, see an image of the back of his head from the point in time $t_1 = e^{-2\pi(1\pm b_0)}t_2$. Note that the causality violation of this spacetime for the case $b_0 > 1$ becomes physically manifest as the observer would see a future image of himself in the direction of $+\eta_1$. This yields the usual causal paradoxes that would occur if, for example, the observer decided to change the outcome of his observation.

The spacetime admits a red-shift potential [HP88], and upon returning, the light signal is subject to a change in frequency f given by

$$\frac{f(t_2)}{f(t_1)} = \frac{a(t_1)}{a(t_2)} = e^{-2\pi(1\pm b_0)}.$$

This means that in the case $b_0 > 1$, the signal coming from the direction of $+\eta_1$ is in fact *blue-shifted*.

In the case $a = a_-$, the time it takes for the light rays emitted by the observer to return is determined by

$$\tanh\left(\frac{t_1}{2\alpha}\right) = e^{-2\pi(1\pm b_0)} \tanh\left(\frac{t_2}{2\alpha}\right).$$

Thus, light from the observer in the direction of $+\eta_1$ only returns if it was emitted during the epoch $t < \alpha \log(\coth(\pi(1 + b_0)))$. Light emitted in the $-\eta_1$ direction meets the observer again if it was emitted at times $t < \alpha \log(\coth(\pi(1 - b_0)))$ provided that $b_0 < 1$. It always returns if $b_0 > 1$, again from a future event on the observer's worldline – and with a blue-shift that increases indefinitely with the observer's proper time.

As for examples of timelike geodesics, in the case $b_0 < 1$ the observer field $W = \cosh(\phi(t))\eta_0 - \sinh(\phi(t))\eta_1$ with

$$\tanh(\phi(t)) = \frac{cb_0a(t)^2 \pm (1 - b_0^2)\sqrt{1 + ca(t)^2}}{ca(t)^2 + 1 - b_0^2}$$

or $\tanh(\phi(t)) = b_0$ has vanishing acceleration. For any $v_0 \neq b_0$ with $-1 < v_0 < 1$, and $t_1 > 0$, we may choose a sign and the constant $c > 0$ such that $\tanh\phi(t_1) = v_0$. Observe that we have $\lim_{t\to\infty} \tanh\phi(t) = b_0$. Thus, a free-falling test particle emitted by the observer in the direction of $-\eta_1$ with arbitrary Newtonian velocity $v = v_0$ accelerates with respect to the fluid in this (or the opposite) direction in order to eventually reach the constant terminal velocity $v = b_0$. Similar to light signals, the test particle may return to the observer, or come arbitrarily close to it.

5.2   The Tilted Closed FLRW Model: Geodesic Flow

Alternatively to the conformally stationary metric discussed in the last section, we may write down a metric where the shift form does not depend on t, i.e., for which the integral curves of the reference frame $V = \partial_t$ are geodesics:

$$g = -(\mathrm{d}t - b_0(r^2\mathrm{d}\mu + (1 - r^2)\mathrm{d}\nu))^2 + a^2(\frac{1}{1 - r^2}\mathrm{d}r^2 + r^2\mathrm{d}\mu^2 + (1 - r^2)\mathrm{d}\nu^2).$$

In this case, the anisotropic pressure with respect to V may again be made to vanish, more precisely if

$$\ddot{a}a - 2(\dot{a})^2 + 2 = 0$$

holds. This equation may be integrated to yield the following solutions:

Proposition 5.2.1. *The geodesic tilted closed FLRW spacetime is a viscous fluid with respect to the standard observer field $V = \partial_t$ if and only if $b_0 = 0$ or the scale parameter is given by either*

$$a_-(t) = \alpha \frac{2\operatorname{sn}(\frac{t-t_0}{\alpha}; \frac{\sqrt{2}}{2})}{1 + \operatorname{cn}(\frac{t-t_0}{\alpha}; \frac{\sqrt{2}}{2})}, \qquad t_0 < t < t_0 + 2K(\frac{\sqrt{2}}{2})\alpha,$$

$$a_0(t) = t - t_0, \qquad t_0 < t,$$

$$a_+(t) = \frac{\sqrt{2}}{\omega}\operatorname{cn}(\omega(t - t_0) - K(\frac{\sqrt{2}}{2}); \frac{\sqrt{2}}{2}), \qquad t_0 < t < t_0 + 2\frac{K(\frac{\sqrt{2}}{2})}{\omega}.$$

Here, sn and cn denote the Jacobi elliptic functions, and K is the complete elliptic integral of the first kind; t_0, α, ω are constants of integration.

Remark 5.2.2. Again, the index properly describes the sign of the deceleration parameter, and we have $\lim_{\alpha\to\infty} a_-(t) = \lim_{\omega\to 0} a_+(t) = a_0(t)$. Also note that a_- becomes infinite in finite time and is thus a Big Rip solution.

Also, we wish to recall the definitions of the elliptic functions used in the proposition above. Fix a real number $0 \le m \le 1$ and compute

$$t(u; m) = \int_0^u \frac{\mathrm{d}\tau}{\sqrt{1 - m^2\sin^2(\tau)}}$$

for arbitrary $u \in \mathbb{R}$. The function t is invertible with respect to u, i.e., we may write $u = u(t; m)$. Define

$$\operatorname{sn}(t; m) = \sin(u(t; m)), \quad \operatorname{cn}(t; m) = \cos(u(t; m)), \quad K(m) = u(\frac{\pi}{2}; m).$$

The matter variables compute to:

$$\rho = a^{-4}(7b_0^2 + 3a^2 + 3(a^2 - b_0^2)(\dot{a})^2),$$
$$p^* = a^{-4}(-b_0^2 + (4b_0^2 - 1)a^2 + (-5a^2 + b_0^2)(\dot{a})^2)$$
$$q = 2a^{-2}((\dot{a})^2 - 3)b.$$

Assuming a vanishing bulk viscosity, i.e., $p = p^*$, we have that $d\rho \wedge dp \neq 0$, thus no equation of state of the form $\rho = \rho(p)$ may hold. For the simplest case $a_0(t) = t$, the spacetime satisfies the dominant and strong energy conditions. Employing the same model assumptions as for the conformally stationary case, we find:

Proposition 5.2.3. *If $a(t) = t$, the geodesic tilted closed FLRW spacetime is a viscous fluid with vanishing bulk viscosity satisfying the laws of linear thermodynamics with*

$$k = \frac{4b_0^2}{T_0}\frac{\sqrt{b_0^2 + t^2}}{t^3}, \quad T = \frac{T_0}{b_0}\sqrt{b_0^2 + t^2}, \quad S = \frac{4b_0^2}{T_0}\frac{\sqrt{b_0^2 + t^2}}{t^4}.$$

Consequently, entropy production is given by

$$\mathrm{div}(SV + \frac{1}{T}q^\sharp) = \frac{4b_0^3}{t^3 T_0(b_0^2 + t^2)^{\frac{3}{2}}}.$$

5.2.1 Geodesics and Observations.

We denote the non-causal and causal regions of the tilted closed FLRW spacetime with geodesic flow with $M_\mathrm{I} = \{(t,p) \in M | a(t) < b_0\}$ and $M_\mathrm{II} = \{(t,p) \in M | a(t) > b_0\}$, respectively. The future-pointing null vector fields

$$\xi_\pm = \pm(a \pm b_0)^{-1}(\eta_0 \pm \eta_1) = a^{-1}((1 \pm b_0)\partial_t \pm a^{-1}(\partial_\mu + \partial_\nu))$$

with $\eta_0 = \partial_t$, $\eta_1 = a^{-1}(b_0\partial_t + \partial_\mu + \partial_\nu)$ are geodesic; ξ_- is only defined on M_I. In the following, we discuss the case $a_0(t) = t$, $b_0 > 0$. The integral curves of above vector fields are given by

$$t(\tau) = \sqrt{\pm 2\tau + t(0)^2}, \quad r(\tau) = \mathrm{const.},$$
$$\mu(\tau(t)) = \nu(\tau(t)) = \pm \log\left(\frac{t(\tau) \pm b_0}{t(0) \pm b_0}\right).$$

This means that in M_I, light rays emitted by a fluid observer in the direction of $-\eta_1$ run into the singularity that lies in his or her past. This fate is not exclusively shared by light rays: The flow of the vector field W_- defines a family of timelike geodesics with $t'(\tau) < 0$, where $W_\pm = \cosh(\phi_\pm(t))\eta_0 \pm \sinh(\phi_\pm(t))\eta_1$ with

$$\phi_\pm(t) = \log\frac{c + \sqrt{c^2 + t^2 - b_0^2}}{b_0 \pm t}, \quad c \geq b_0.$$

Also observe that for any choice of the constant c, $\phi_-(t) \geq \log((b_0 - t)^{-1}(b_0 + t))$ holds, which shows that any test particle emitted by a fluid observer at proper time t_1 with Newtonian velocity $v_0 = \tanh \phi_-(t_1) \geq \frac{2b_0 t_1}{b_0^2 + t_1^2}$ in the direction of $-\eta_1$ ends up in the singularity $\{t = 0\}$, as well.

In contrast to this, the t-coordinate increases along the integral curves of W_+, and for any starting velocity v_0 at the fluid observer's proper time t_1 we may find a parameter c such that $\tanh \phi_+(t_1) = v_0$. Thus, test particles emitted in the direction of $+\eta_1$ simply pass the causal horizon $\{t = b_0\}$ (just as the fluid matter itself).

Also, though possessing no past endpoint, the null geodesics in the direction of ξ_- are totally past-imprisoned in a compact set $[b_0 - \epsilon, b_0] \times S^3$. Similarly, future-directed null geodesics contained in the causal region M_{II} in the direction of $\xi_-^{II} = (a - b_0)^{-1}(\eta_0 - \eta_1)$ are past-imprisoned in a compact set $[b_0, b_0 + \epsilon] \times S^3$. Light signals emitted in the direction of $+\eta_1$, however, are not affected by the causal horizon. This means that a fluid observer in M_{II} sees the singularity in the direction of $-\eta_1$ but only sees light emitted by the causal horizon in the direction of $+\eta_1$.

Remark 5.2.4. The existence of a totally imprisoned non-spacelike curve was characterized with [Min09b, Theorem 1] as equivalent to the existence of a non-stably causal non-empty open subset with compact closure. In our case, $]0, b_0[\times S^3$ would be an example of such a subset.

5.3 A Causality Violating Stationary Perfect Fluid Solution

The Hopf vector field $K = -b_0(\partial_\mu + \partial_\nu)$ on S^3 is a geodesic Killing vector field. From Proposition 4.4.2 it follows that in the case $b_0 > 1$, the lift of K to the conformally stationary tilted closed FLRW spacetime (M, g) defined in Section 5.1 is a future-pointing timelike Killing vector field. We may thus investigate the different cosmological model (M, g, W) with stationary observer field, $W = (-g(\hat{K}, \hat{K}))^{-\frac{1}{2}}\hat{K}$.

Provided that the scale parameter is one of the following:

$$a_-(t) = k\alpha \sinh\left(\frac{t - t_0}{\alpha}\right), \qquad\qquad t_0 < t,$$

$$a_0(t) = k(t - t_0), \qquad\qquad t_0 < t,$$

$$a_+(t) = \frac{k}{\omega}\sin(\omega(t - t_0)), \qquad\qquad t_0 < t < t_0 + \frac{\pi}{\omega},$$

with $k = \sqrt{\frac{b_0^2+1}{b_0^2-1}}$, it is in fact of perfect fluid type with

$$\rho_W = a^{-2}\left(-3(\dot{a})^2(b_0^2-1) + 5b_0^2 + 3\right),$$
$$p_W^* = a^{-2}\left(3(\dot{a})^2(b_0^2-1) - 3b_0^2 - 3\right).$$

Note that ρ_W and p_W^* are constant along integral curves of W since $g(\hat{K}, \nabla t) = \bar{b}^i t_i = 0$.

5.4 Barotropic Perfect Fluid Solutions with Geodesic Flow

We now wish to classify tilted warped product spacetimes with proper geodesic perfect fluid source, i.e., we have $\dot{b}_i = 0$ and

$$R_{ij} - \frac{R}{2}g_{ij} = (\rho + p)u_i u_j + pg_{ij},$$

where $\rho + p \neq 0$. Furthermore, we require the fluid to be barotropic, i.e., an equation of state $\rho = \rho(p)$ with $\frac{d\rho}{dp} \neq 0$ holds. Physically, models with vanishing acceleration and shear are important since they follow an isotropic Hubble law of first order [HP99].

Proposition 5.4.1. *For any 4-dimensional tilted warped product spacetime (M, g) with barotropic geodesic perfect fluid source, $\rho + p \neq 0$, either one of the two following cases may occur:*

1. *The vorticity vanishes. In this case, (M, g) is locally isometric to an FLRW spacetime, and the fiber (N, h) is locally conformally equivalent to a space of constant curvature. This holds globally if the first Betti number of M vanishes.*

2. *The expansion vanishes. In this case, the following holds:*
   (a) The pressure p, energy density ρ and rotation scalar $|\omega|^2$ are constant.
   (b) If $|\omega| \neq 0$, (M, g) is locally isometric to the classical Gödel spacetime.
   (c) If $|\omega| = 0$ and $\rho > 0$, (M, g) is locally isometric to Einstein's static universe.
   These results hold globally if (M, g) is complete and simply connected.

Proof. In the following we make frequent use of Corollary 4.3.3. By FLRW spacetime we mean a warped product $(I' \times N, -dt'^2 + \tilde{h})$ with a fiber $(N, \tilde{h})$ of constant curvature (which in $d - 1 = 3$ dimensions is equivalent to $(N, \tilde{h})$ being Einstein). One may also require the fiber to be complete in which case the spacetime is not only stably causal but globally hyperbolic [Sán99]. Now, non-accelerating shear-free barotropic perfect fluids cannot both be rotating and expanding [SSS98]. Let us first examine the case of vanishing vorticity. In this case, at least locally we must have $b = df$ for some function $f\colon N \to \mathbb{R}$ (globally if we assume $B_1(M) = B_1(N) = 0$). Furthermore, the heat flux vanishes if and only if $\dot\theta = 0$ or $b_i = 0$. If the shift vanishes, $b_i = 0$, we are finished since the vanishing of the anisotropic pressure immediately yields $r_{ij} = \frac{\alpha}{3}h_{ij}$. Thus, $a(t) = ce^{\theta t}$ for some constants $c > 0$ and θ. If we pull back the metric

$$g = -(\mathrm{d}t - \mathrm{d}f)^2 + c^2 e^{2\theta t} h$$

via the change of slicing $(t', x) = (t - f(x), x)$ we see that it becomes a warped product:

$$\Phi^* g = -\mathrm{d}t'^2 + c^2 e^{2\theta t'} e^{2\theta f} h.$$

The fiber metric $\tilde h = e^{2\theta f} h$ is conformally equivalent to h; its Ricci tensor may be computed via the formula from Proposition 2.3.6:

$$\tilde r_{ij} = r_{ij} - \theta \mathrm{D}_i f_{,j} + \theta^2 f_{,i} f_{,j} - (\ldots)h_{ij}$$
$$= r_{ij} - \theta \mathrm{D}_{(i} b_{j)} + \theta^2 b_i b_j - (\ldots)h_{ij}.$$

Plugging this into the formula for the anisotropic pressure, we have $0 = \Pi_{ij} = \tilde r_{ij} - \frac{\tilde\alpha}{3}\tilde h_{ij}$ as desired.

For the second case of vanishing expansion ($a = 1$ without loss of generality), the conditions of a vanishing heat flux and anisotropic pressure yield

$$\rho = \tfrac{1}{2}r + \tfrac{3}{2}|\omega|^2, \quad p = -\tfrac{1}{6}r + \tfrac{1}{6}|\omega|^2.$$

It is clear from these formulas and momentum conservation ($h^m{}_j p_{,m} = -(\rho+p)\dot b_j = 0$) that the gradient of p vanishes. The equation of state then implies that ρ and therefore r and $|\omega|^2$ are constant, too. Therefore, Proposition 6.2.11 and Remark 6.2.12 apply, see Section 6.2.1 below.

$\square$

Since shear-free dusts also either expand or rotate [Ell67], we may state by very similar arguments:

Proposition 5.4.2. *For any tilted warped product spacetime (M, g) with geodesic dust source, either one of the two following cases may occur:*

1. *The vorticity vanishes. In this case, (M, g) is an FLRW dust solution.*

2. *The expansion vanishes. In this case, the vorticity also vanishes if and only if the spacetime is flat.*

Example 5.4.3. Let N be $\mathbb{R}^3$ with deleted z-axis. The van Stockum spacetime [Lan23, vS37] $M = \mathbb{R} \times N$ with metric

$$g = -(\mathrm{d}t - \tfrac{1}{\sqrt 2}\alpha r^2 \mathrm{d}\phi)^2 + r^2 \mathrm{d}\phi^2 + e^{-\frac{1}{2}\alpha^2 r^2}(\mathrm{d}r^2 + \mathrm{d}z^2),$$

where (r, ϕ, z) are cylindrical coordinates on N, and $\alpha \in \mathbb{R}$, is a rotating dust solution covered by Proposition 5.4.2, with non-constant energy density $\rho = 2\alpha^2 \exp(\frac{1}{2}\alpha^2 r^2)$. Note that $B_1(N) \neq 0$.

Remark 5.4.4. Note that in recent physical cosmology, models for the observable universe are discussed the spatial part of which is homeomorphic to the Poincaré homology sphere [LWR+03, LR08]. This is the unique Poincaré space with finite fundamental group, in particular it is not simply connected but has vanishing first Betti number.

Chapter 6

GLOBAL ASPECTS OF STATIONARY SPACETIMES

In this chapter, we investigate stationary spacetimes in a standard product form in more detail, with an emphasis on methods with a global flavor.

6.1 Causality and Sobolev Inequalities

This section continues from Section 4.2. Many results may be given analogously for general tilted products by investigating the shift form on slices for fixed values of the standard observers' proper time t.

In the following, let (N, h) be a connected Riemannian manifold, $\dim(N) \geq 1$, $\xi \in \Gamma(TN^*)$ and $\beta \in \mathfrak{F}(N)$ with $\beta > 0$. Consider the stationary metric

$$g = -(\mathrm{d}t + \xi)^2 + \beta^{-2}h \tag{6.1}$$

defined on $\mathbb{R} \times N$. This spacetime is stably causal if $|\xi|_h < \beta^{-1}$ holds everywhere on N. This is true, in particular, if

$$\|\xi\|_{L^\infty} = \sup(|\xi|_h) < \inf(\beta^{-1}) = \|\beta\|_{L^\infty}^{-1} \tag{6.2}$$

holds (Proposition 4.2.2). However, the metric g is invariant with respect to any change of slicing $\xi \mapsto \xi' = \xi + \mathrm{d}f$. Conversely, conformally transforming the metric (6.1) by multiplying with β^2 immediately yields the following interesting fact:

Proposition 6.1.1. *If the spacetime $(\mathbb{R} \times N, g)$ with*

$$g = -\beta^2(\mathrm{d}t + \xi)^2 + h$$

is non-stably causal, then for any function $f \in \mathfrak{F}(N)$,

$$\|\xi - \mathrm{d}f\|_{L^\infty} \geq \|\beta\|_{L^\infty}^{-1}.$$

It is natural to ask for conditions on a given shift form ξ, not necessarily with $|\xi| < \beta^{-1}$, which imply that there exists a representative $\xi' = \xi + \mathrm{d}f$ with $|\xi'| < \beta^{-1}$. Intuitively, one would expect that such a change of slicing exists if ξ is close to an exact form $\mathrm{d}f$, and that the failure to be exact is measured by $|\mathrm{d}\xi|$ if $B_1(N) = 0$. For compact fibers we have indeed:

Proposition 6.1.2. *Let (N, h) be compact and orientable, and $q \in \mathbb{R} \cup \{\infty\}$ with $q > \dim(N)$. Then there exist numbers $C_1 \geq 0$ and $C_2 \geq 0$, depending only on q and (N, h), such that the inequality*

$$C_1 \|\xi\|_{L^q} + C_2 \|\mathrm{d}\xi\|_{L^q} < \|\beta\|_{L^\infty}^{-1} \tag{6.3}$$

implies that the stationary spacetime $(\mathbb{R} \times N, g)$ with

$$g = -\beta^2 (\mathrm{d}t + \xi)^2 + h$$

is stably causal. If $q = \infty$, we may set $C_1 = 1$ and $C_2 = 0$. If the first Betti number of N vanishes, we may set $C_1 = 0$.

Proof. First, we may again consider the conformally equivalent metric (6.1). The case $q = \infty$ is given by inequality (6.2). Then we mimic the proof of Theorem 1.1 in [GT06]: Denote by H the $\langle\!\langle \cdot, \cdot \rangle\!\rangle$-orthogonal projection of ξ onto the space of harmonic one-forms $H^1(N)$, whereas G is the Green operator of the elliptic equation $\triangle \mu = \xi - H\xi$ for $\mu \in \Lambda^1(N)$. For details on these operators, see [dR84, §31].

Since $\triangle G = \mathrm{id} - H$ and $\mathrm{d}G = G\mathrm{d}$, we have

$$\mathrm{d}\delta G = \triangle G - \delta \mathrm{d}G = \mathrm{id} - H - \delta G \mathrm{d}.$$

Since δG and H are bounded linear operators from the space $L_1^q(N)$ to the space $W_1^{1,q}(N)$, for the exact one-form $\eta = \mathrm{d}\delta G\xi$,

$$\|\xi - \eta\|_{W^{1,q}} = \|H\xi + \delta G \mathrm{d}\xi\|_{W^{1,q}} \leq c_1 \|\xi\|_{L^q} + c_2 \|\mathrm{d}\xi\|_{L^q}.$$

Note that $c_1 = 0$ holds if and only if H is the zero operator if and only if $\dim(H^1(N)) = B_1(N) = 0$. Because of the Sobolev-type inequalities for differential forms, we have

$$\|\tau\|_{L^\infty} \leq c \|\tau\|_{W^{1,q}}$$

for every differential form τ provided that $q > \dim(N)$. With the remarks above, the result follows with $C_1 := cc_1$ and $C_2 := cc_2$. $\qquad\square$

In the following, (N, h) is always compact, orientable and with vanishing first Betti number. Note that this means $\dim(N) \geq 2$, and N is diffeomorphic to the sphere for $\dim(N) = 2$. For any given $q \in \mathbb{R} \cup \{\infty\}$ with $q > \dim(N)$, the proposition above shows that for some number $C = C(q)$, $\|\mathrm{d}\xi\|_{L^q} < C(q) \|\beta\|_{L^\infty}^{-1}$ implies that $g = -\beta^2 (\mathrm{d}t + \xi)^2 + h$ is stably causal. In the following, we discuss ways to obtain an upper bound for $C(q)$. This is important since such a bound would yield an explicit sufficient condition on β and ξ—that is independent of the slicing—for the stable causality of the corresponding stationary spacetime.

For a unit vector field K on (N, h), write $\xi = h(K, \cdot)$, and note that the corresponding stationary spacetime with $\beta = 1$ is non-stably causal by Proposition 4.2.2. Hence,

$$C(q) \leq \|\mathrm{d}\xi\|_{L^q} = \|\operatorname{curl}(K)\|_{L^q}. \tag{6.4}$$

This holds for any unit vector field, thus

$$C(q) \leq \inf_{K \in \Gamma(TN), |K|=1} \|\operatorname{curl}(K)\|_{L^q}.$$

This inequality is of course not optimal; for manifolds with non-vanishing Euler characteristic, for example, it yields no finite bound at all since every vector field K must have a critical point $p \in N$ where $K_p = 0$.

To specialize further, suppose (N, h) admits a unit Killing vector field K. Note that this means that $\dim(N) \geq 3$ since the 2-sphere has Euler characteristic $\chi(S^2) = 2$. Concerning such vector fields, note the following (see [BN06, BN08] for details):

• We have that K is a geodesic vector field: for any $X \in \Gamma(TN)$,

$$0 = X(h(K, K)) = 2h(\mathrm{D}_X K, K) = -2h(\mathrm{D}_K K, X),$$

where D is the Levi-Civita connection of (N, h). Conversely, if a non-trivial Killing vector field is geodesic, it may be rescaled by a constant to yield a unit Killing vector field.

• The curl of K must be non-zero since any complete Riemannian manifold admitting a gradient field of constant length is non-compact [Sak96]. This implies that the Ricci curvature ric of (N, h) cannot be non-positive. More precisely, because of Raychaudhuri's equation:

$$0 \leq \frac{1}{4}|\operatorname{curl}(K)|^2 = \operatorname{ric}(K, K), \tag{6.5}$$

with $\operatorname{ric}(K, K) > 0$ somewhere.

• Even-dimensional compact Riemannian manifolds with positive sectional curvature do not admit a unit Killing vector field.

• Any three-dimensional compact simply connected manifold is diffeomorphic to the 3-sphere and can therefore be furnished with a Riemannian metric that admits a unit Killing vector field.

• Important examples of unit Killing vector fields are *Sasakian structures* [Sas60, Str10]. A Sasakian structure is a unit Killing vector field K such that

$$\mathrm{D}_X(\mathrm{d}K^\flat) = 2K^\flat \wedge X^\flat$$

holds for all $X \in \Gamma(TN)$.

The inequality (6.4) may be written in terms of Ricci curvature using Raychaudhuri's equation (6.5). Obviously, the most simple cases are those where the function $\mathrm{ric}(K, K)$ is a constant. This is the case, for example, if (N, h) is Einstein, or K is a Sasakian structure:

Proposition 6.1.3. *Let $D = \dim(N)$ and suppose that (N, h) admits a unit Killing vector field K.*

1. *If (N, h) is Einstein,*

$$C(q) \le 2\sqrt{\frac{\kappa^h}{D}} \cdot \mathrm{vol}(N)^{\frac{1}{q}}.$$

2. *If K is a Sasakian structure,*

$$C(q) \le 2\sqrt{D-1} \cdot \mathrm{vol}(N)^{\frac{1}{q}}.$$

Proof. 1. Since (N, h) is Einstein, we have

$$|\mathrm{curl}(K)| = 2(\mathrm{ric}(K, K))^{\frac{1}{2}} = 2(D^{-1}\kappa^h)^{\frac{1}{2}} = \mathrm{const.},$$

therefore

$$C(q) \le \left(\int_N |\mathrm{curl}(K)|^q \, \mathrm{dvol}(h) \right)^{\frac{1}{q}} = 2(D^{-1}\kappa^h)^{\frac{1}{2}} \cdot \mathrm{vol}(N)^{\frac{1}{q}}$$

2. If K is Sasakian, $\mathrm{ric}(K, K) = D - 1$, see [Str10, Proposition 1.1.4]. □

Example 6.1.4. Fix $m \in \mathbb{N}$, $m \ge 1$, and let $K \colon S^{2m+1} \to TS^{2m+1}$ be a unit Hopf vector field on the round unit $2m + 1$-sphere. If we picture S^{2m+1} as a submanifold of $\mathbb{R}^{2m+2}$, such a vector field is given by $K(x) = Jx$ where J is the standard complex structure on $\mathbb{C}^{m+1} \cong \mathbb{R}^{2m+2}$, or, more explicitly

$$K(x_1, \ldots, x_{2m+2}) = (x_{m+2}, \ldots, x_{2m+2}, -x_1, \ldots, -x_{m+1})^T.$$

We have that K is a Sasakian structure on a simply connected manifold, thus, a bound for the odd-dimensional unit sphere is given by $(q > 2m + 1)$:

$$C(q) \le 2\sqrt{2m} \cdot \mathrm{vol}(S^{2m+1})^{\frac{1}{q}} = 2\sqrt{2m} \left(\frac{2\pi^{m+1}}{m!} \right)^{\frac{1}{q}}.$$

Example 6.1.5. Let $\epsilon > 0$ and (S^3, h_ϵ) be a Berger sphere, i.e., $h_\epsilon = h_1 + (\epsilon^2 - 1)h_1(B, \cdot) \otimes h_1(B, \cdot)$ where h_1 is the round metric on S^3 and B is a Hopf vector field with $h_1(B, B) = 1$. We have that $K := \epsilon^{-1}B$ is an h_ϵ-unit Killing vector field. For $\epsilon \ne 1$, it is neither a Sasakian structure, nor is (S^3, h_ϵ) an Einstein manifold.

Nevertheless, the Raychaudhuri scalar with respect to K is a constant, and one easily computes

$$C(q) \leq 2\sqrt{2}\epsilon \left(2\pi^2\epsilon\right)^{\frac{1}{q}}$$

for (S^3, h_ϵ), $q > 3$.

Example 6.1.6. A more sophisticated example is given by a metric h defined on $N = S^2 \times S^3$ that may be written as follows [GMSW04]:

$$h = \frac{1}{6}(d\theta^2 + \sin^2\theta d\phi^2 + d\alpha^2 + \sin^2\alpha d\nu^2) + \frac{1}{9}(d\psi - \cos\alpha d\phi - \cos\alpha d\nu)^2,$$

where the manifold is covered by a parametrization with range $0 \leq \theta \leq \pi$, $0 \leq \phi \leq 2\pi$, $0 \leq \psi \leq 4\pi$, $0 \leq \nu \leq 2\pi$, $0 \leq \alpha \leq \frac{\pi}{2}$. We have that (N, h) is an Einstein space, and $K = 3\partial_\psi$ is a Sasakian structure. It is therefore easy to check that

$$C(q) \leq 4(\text{vol}^h(N))^{\frac{1}{q}} = 4\left(\frac{8}{27}\pi^3\right)^{\frac{1}{q}}$$

for $q > 5$.

Proposition 6.1.7. Let K be a conformal vector field on (N, h) with no critical points, i.e., $|K| > 0$ holds everywhere. Then

$$C(q) \leq \| \operatorname{curl}(|K|^{-2}K)\|_{L^q}\|K\|_{L^\infty}.$$

Proof. Write $\xi = h(K, \cdot)$. We have that K is a unit Killing vector field on $(N, \tilde{h})$ with $\tilde{h} = |K|^{-2}h$. Define $\tilde{\xi} = \tilde{h}(K, \cdot)$ and apply the above arguments to the spacetime

$$g = -(dt + \tilde{\xi})^2 + \tilde{h} = -(dt + |K|^{-2}\xi)^2 + |K|^2 h.$$

$\square$

6.2 Charged Perfect Fluid Solutions of Gödel-type

The study of so-called Gödel-type spacetimes has a long history [PS10]. There is no general agreement to the precise meaning of the term "Gödel-type"; however, Definition 6.2.3 below covers a number of important solutions discovered over the years such as, for example, the classical Gödel spacetime, the solution found in [BB68], and the class (a) of solutions given in [Upo94]. The goal of this section is to show some curvature and structure properties of these spacetimes, and to classify all four-dimensional models with constant vorticity scalar.

Definition 6.2.1. An *electromagnetic field tensor* F on a spacetime (M, g) is formally a closed two-form on M. The energy–momentum tensor associated with F is given by

$$\tau = \tau^F = -F \circ F - \frac{1}{4}\langle F, F\rangle g,$$

where $(F \circ F)_{ij} := F_i{}^k F_{kj}$.

Remark 6.2.2. The requirement $\mathrm{d}F = 0$ is the homogeneous part of Maxwell's equations, the inhomogeneous part being $\delta F = sV^\flat$ where V is the reference frame comoving with the charged material, and $s \in \mathfrak{F}(M)$ is the charge density. Recall that δ denotes the divergence. The energy–momentum tensor may be obtained, for example, from a variational formulation of the Einstein–Maxwell equations [Thi97, Section 10.2].

In the following, let (N, h) be a connected Riemannian manifold, $d = \dim(N) + 1 \geq 2$, and $\xi \in \Gamma(TN^*)$.

Definition 6.2.3. We call the stationary spacetime $(M, g) = (\mathbb{R}, -\mathrm{d}t^2) \times^{\pi_2^*\xi} (N, h)$ a *charged perfect fluid of Gödel type* if the Einstein–Hilbert equation for a charged perfect fluid with fluid vector field $V = \partial_t$ and electromagnetic field tensor $F = \frac{k}{2}\mathrm{d}V^\flat = \frac{k}{2}\pi_2^*\mathrm{d}\xi$ hold for some number $k \in \mathbb{R}$, i.e.,

$$\mathrm{Ric} - \frac{\kappa}{2}g = \rho V^\flat \otimes V^\flat + ph + \tau^F$$

for functions $\rho, p \in \mathfrak{F}(M)$.

For the rest of this section, we assume that $(M, g) = (\mathbb{R}, -\mathrm{d}t^2) \times^\xi (N, h)$ is a charged perfect fluid spacetime of Gödel type, and $\Omega := -\frac{1}{2}\mathrm{d}\xi$.

Proposition 6.2.4. *The following statements hold:*

1. *The divergence of Ω vanishes. If N is orientable and Ω has compact support, Ω necessarily vanishes. Thus, (M, g) must be static in this case.*

2. *The energy density and the pressure are given by:*

$$\rho = \left(\frac{3}{2} - \frac{k^2}{4}\right)|\Omega|^2 + \frac{\kappa^h}{2}, \tag{6.6}$$

$$p = \frac{1}{4(d-1)}((d-5)(k^2-2)|\Omega|^2 - 2(d-3)\kappa^h).$$

3. *We have that $\delta F = k|\Omega|^2 V^\flat$, i.e., the inhomogeneous Maxwell equation is satisfied, and the charge density is given by $s = k|\Omega|^2$. The electromagnetic field is purely magnetic, i.e., $F(V, \cdot) = 0$.*

Proof. Formulas for the Ricci curvature of (standard) stationary spacetimes are well-established [Ger71]. Via the substitution $b = \mathrm{d}(\log \beta) - t\beta\xi$, here $\beta = 1$, we may also use Proposition 4.3.1. Omitting overbars, a direct computation shows that the matter part of the spacetime's energy–momentum is given by

$$R_{ij} - \frac{R}{2}g_{ij} - \tau_{ij} = r_{ij} + (k^2 - 2)\Omega^k{}_j\Omega_{ik} + 2\mathrm{D}^k\Omega_{k(i}u_{j)} \tag{6.7}$$

$$+ \left(\left(\frac{3}{2} - \frac{k^2}{4}\right)|\Omega|^2 + \frac{r}{2}\right)u_iu_j + \left(-\frac{r}{2} + \left(\frac{k^2}{4} - \frac{1}{2}\right)|\Omega|^2\right)h_{ij}.$$

1. The vanishing of the heat flux enforces $\mathrm{D}^k\Omega_{ki} = 0$, i.e., $\delta\Omega = 0$. Using Green's identity (Theorem 2.2.40), we have

$$\|\Omega\|^2_{L^2} = \langle\!\langle \Omega, \Omega\rangle\!\rangle = -\frac{1}{2}\langle\!\langle \Omega, \mathrm{d}\xi\rangle\!\rangle = \frac{1}{2}\langle\!\langle \delta\Omega, \xi\rangle\!\rangle = 0.$$

2. The vanishing of the anisotropic pressure is equivalent to

$$r_{ij} + (k^2 - 2)\Omega^k{}_j\Omega_{ik} = \frac{\alpha}{d-1}h_{ij} \tag{6.8}$$

for some function $\alpha \in \mathfrak{F}(N)$. This function may be computed via the trace: $\alpha = r - (k^2 - 2)|\Omega|^2$. Substituting into Eq. (6.7) yields the result.

3. It is obvious that $F(V, \cdot) = 0$ since (the lift of) Ω is a horizontal tensor field. The inhomogeneous Maxwell equation is the consequence of a straightforward calculation using the formulas established in the proof of Proposition 4.3.1:

$$\Omega^k{}_{i;k} = \Omega^k{}_{i\,k} - \Gamma^l{}_{ik}\Omega^k{}_l + \Gamma^k{}_{lk}\Omega^l{}_i = \Omega^k{}_{i|k} - \beta^l{}_{ik}\Omega^k{}_l + \beta^k{}_{lk}\Omega^l{}_i$$

$$= -\beta^l{}_{ik}\Omega^k{}_l = \frac{1}{2}(u_if^l{}_k + u_kf^l{}_i)\Omega^k{}_l = \Omega_k{}^l\Omega^k{}_lu_i.$$

$\square$

Remark 6.2.5. The solutions discussed in [GKS05] present the case $k^2 = 2$. Cf. also [Gür10, Proposition 2]. Of course, the case $k = 0$ corresponds to perfect fluids with vanishing charge.

Proposition 6.2.6. *The following statements hold:*

1. *If $d \neq 3$ and $k^2 = 2$ hold, (N, h) is an Einstein space.*

2. *If (N, h) is an Einstein space and $k^2 \neq 2$, then $\Omega = 0$ or the spacetime dimension d is odd.*

Proof. 1. This is an immediate consequence of Eq. (6.8).

2. Set up an orthonormal frame. If $d - 1$ is odd, the matrix (Ω_{ik}) cannot be a non-trivial square root of a multiple of the identity matrix (h_{ij}); it is impossible for $(\Omega^k{}_j \Omega_{ik})$ to have maximal rank.

$\square$

Proposition 6.2.7. *The following statements hold:*

1. *If $d = 3$ or (N, h) is a space of constant scalar curvature, $|\Omega|^2 = $ const.*
2. *If $d = 5$ or $|\Omega|^2 = $ const., (N, h) is a space of constant scalar curvature.*

Proof. Since the divergence and exterior derivative of Ω vanish,

$$\mathrm{D}^l(\Omega^k{}_l \Omega_{ik}) = (\mathrm{D}^l \Omega^k{}_l)\Omega_{ik} + \Omega^{kl}\mathrm{D}_l\Omega_{ik} = \Omega^{kl}\mathrm{D}_l\Omega_{ik}$$
$$= \Omega^{kl}(-\mathrm{D}_k\Omega_{li} - \mathrm{D}_i\Omega_{kl}) = -\frac{1}{2}\mathrm{D}_i|\Omega|^2 - \Omega^{kl}\mathrm{D}_l\Omega_{ik},$$

i.e., $\mathrm{D}^l(\Omega^k{}_l \Omega_{ik}) = -\frac{1}{4}\mathrm{D}_i|\Omega|^2$. Substituting this into the divergence of Eq. (6.8), and using the contracted Bianchi identity $\mathrm{D}^k r_{ki} = \frac{\mathrm{D}_i r}{2}$, we conclude

$$2(d - 3)\mathrm{D}_i r = (k^2 - 2)(d - 5)\mathrm{D}_i|\Omega|^2.$$

$\square$

Remark 6.2.8. It is clear that ρ and p are constant whenever $|\Omega|^2$ and κ^h are constant because of Eq. (6.6). The converse is true if $k^2 \neq 2(d - 2)$.

Proposition 6.2.9. *Suppose that $d \neq 3$ and Ω is covariantly constant, i.e., $\mathrm{D}\Omega = 0$. Then, (N, h) is locally isometric to a Riemannian product of Einstein spaces. If (N, h) is simply connected and complete, this holds globally.*

Proof. To show that (N, h) is a product of Einstein spaces, by the de Rham–Wu decomposition theorem, it suffices to show that $\mathrm{D}_k r_{ij} = 0$; see [BB01], for example. Suppose $\mathrm{D}_k\Omega_{ij} = 0$. By covariantly derivating Eq. (6.8) and using the contracted Bianchi identity $\mathrm{D}^k r_{kj} = \frac{\mathrm{D}_j r}{2}$, the result follows. $\square$

Remark 6.2.10. Note that because of the homogeneous Maxwell equation, the two-form Ω is covariantly constant if and only if it is Killing, i.e., $(\mathrm{D}_X\Omega)(Y, Z) + (\mathrm{D}_Y\Omega)(X, Z) + (\mathrm{D}_Z\Omega)(X, Y) = 0$ holds for all $X, Y, Z \in \Gamma(TN)$.

6.2.1 Four-Dimensional Gödel-type Spacetimes with Constant Vorticity Scalar.

In this subsection, we always assume $d = 4$. In this case, the field equations may be written as follows. The two-form Ω is the Hodge dual of a one-form η: $\Omega = *\eta$, hence

$|\Omega|^2 = \frac{1}{2}|\eta|^2$, and this one-form is closed since Ω is co-closed. Thus, at least locally, there exists a harmonic function f with $\mathrm{d}f = \frac{1}{\sqrt{2}}\eta$. If the first Betti number of N vanishes, then f can be defined globally. This is the case, for example, if N is simply connected. Substituting in Eq. (6.8) yields

$$\mathrm{ric} = -\frac{k^2-2}{2}\mathrm{d}f \otimes \mathrm{d}f + \frac{\alpha_1}{3}h \qquad (6.9)$$

with $\alpha_1 = \kappa^h + \frac{k^2-2}{2}|\mathrm{d}f|^2$.

Proposition 6.2.11. *Suppose (N, h) is complete and simply connected. Also assume that the vorticity scalar $|\Omega|$ is constant and non-zero. Then, (N, h) is isometric to one of the following homogeneous 3-dimensional Riemannian manifolds:*

1. Euclidean space or a hyperbolic space.

2. The Riemannian product of the real line with a 2-sphere or a hyperbolic plane.

3. A particular Sol geometry: $\mathbb{R}^3$ with metric

$$h = \lambda e^{-2z}\mathrm{d}x^2 + \lambda e^{2z}\mathrm{d}y^2 + \frac{1}{\lambda^2}\mathrm{d}z^2, \qquad (6.10)$$

where $\lambda \in \mathbb{R}$ with $\lambda > 0$.

Proof. If $k^2 = 2$, then (N, h) is an Einstein space (Proposition 6.2.6), and therefore a space form since N is complete and simply connected, and $\dim(N) = 3$. Since Ω does not vanish, N cannot be the sphere (Proposition 6.2.4). Therefore, in this case N is either Euclidean space or a hyperbolic space.

Suppose $k^2 \neq 2$. Note that (N, h) has constant principal Ricci curvatures, i.e., constant eigenvalues $\rho_1 = \rho_2 \neq \rho_3$ of the Ricci operator $X \mapsto \mathrm{Ric}(X, \cdot)^\sharp$:

$$\beta := \rho_1 = \rho_2 = \frac{\kappa^h}{3} + \frac{k^2-2}{6}|\Omega|^2,$$

$$\alpha := \rho_3 = \frac{\kappa^h}{3} - \frac{k^2-2}{3}|\Omega|^2.$$

A unit eigenvector field corresponding to the simple eigenvalue is given by $E_3 = |\mathrm{D}f|^{-1}\mathrm{D}f$. The vorticity ω of this vector field vanishes, and Theorem 1 in [Bue96] implies $\alpha = -2|\sigma|^2$ where σ is the shear of E_3. Therefore, $|\sigma|$ is constant and it follows from Theorem 2 in [Bue96] that (N, h) is locally homogeneous. We conclude that N is a non-compact and simply connected, homogeneous space, see [Tri92].

Now, all 3-dimensional, simply connected, homogeneous spaces are in fact explicitly known – a complete list is given in [Pat96], for example. Given $\omega = 0$ and $\sigma = \mathrm{const.}$, it is easy to conclude from the formulas given in [Bue96] that $\alpha \leq 0$, and that $\alpha < 0$

Homogeneous manifold	Signature of the Ricci quadratic form
$\mathbb{R} \times H^2$, $\mathbb{R} \times S^2$	$(-,-,0)$, $(+,+,0)$
Sol^3	$(0,0,-)$, $(-,-,+)$
$\widetilde{SL}(2,\mathbb{R})$	$(0,0,-)$, $(-,-,+)$
$\widetilde{ISO}(2)$	$(-,-,+)$
Nil^3	$(-,-,+)$
Non-unimodular symmetry group	$(-,-,+)$, $(-,-,-)$, $(-,-,0)$

Table 6.1: List of all non-compact, simply connected, homogeneous 3-manifolds with non-constant curvature

implies $\beta = 0$. Therefore, the only possible signatures for the Ricci quadratic form are $(\pm, \pm, 0)$ and $(0,0,-)$. We can exclude $\widetilde{ISO}(2)$ and Nil^3 since the signatures of the Ricci quadratic form do not fit, see Table 6.1 and cf. [Mil76]. The case $(N,h) = \widetilde{SL}(2,\mathbb{R})$ can also be excluded since it corresponds to Bianchi type VIII and would imply $\omega \neq 0$ (see [McM95, Table III]). In fact, the only possible Bianchi type left is VI_0 which corresponds to the Sol geometry. The Sol geometries with Ricci signature $(0,0,-)$ are precisely the ones given by (6.10).

□

All solutions for each class are explicitly listed in Table 6.2, with the exception of the case $(N,h) = H^3$. Given the form of the metric, the solutions for f are unique up to adding a constant, and up to sign. They are implied directly by investigating the Einstein–Hilbert equation with the requirement that f is harmonic and has a gradient of constant norm.

Remark 6.2.12. On complete manifolds with non-negative Ricci curvature, the only functions with a gradient of constant norm are the affine functions, i.e., with $\mathrm{D}\mathrm{d}f = 0$ [Sak96]. The only solution in Table 6.2 where f is not an affine function is given by $(N,h) = \mathrm{Sol}^3$. Also note that the only non-charged ($k = 0$) perfect fluid in the list is the classical Gödel spacetime. From this observation, one may easily conclude Gödel's Theorem [Göd49, Ozs65]: The Gödel spacetime and Einstein's static universe are the only (four-dimensional) stationary Λ-dust solutions to Einstein's equations with positive energy density that are spatially homogeneous.

Fiber	Metric	Gen. function/ vorticity/coupling	Energy dens./ pressure
E^3	$h = \mathrm{d}x^2 + \mathrm{d}y^2 + \mathrm{d}z^2$	$f = \frac{\omega_0}{\sqrt{3}}(x+y+z)$, $\|\Omega\|^2 = \omega_0^2$, $k^2 = 2$	$\rho = \omega_0^2$, $p = 0$
$\mathbb{R} \times S^2$	$h = R\mathrm{d}x^2 + R\sin^2(x)\mathrm{d}y^2$ $+\mathrm{d}z^2$	$f = \sqrt{\frac{2}{k^2-2}}R^{-1}z$, $\|\Omega\|^2 = \frac{2}{k^2-2}R^{-2}$, $k^2 > 2$	$\rho = \frac{k^2+2}{2(2-k^2)}R^{-2}$, $p = -\frac{1}{2}R^{-2}$
$\mathbb{R} \times H^2$	$h = \mathrm{d}x^2 + \frac{1}{2}\mathrm{e}^{2\sqrt{2}\omega_0 x}\mathrm{d}y^2$ $+\mathrm{d}z^2$	$f = \frac{2\omega_0}{\sqrt{2-k^2}}z$, $\|\Omega\|^2 = \frac{4\omega_0^2}{2-k^2}$, $k^2 < 2$	$\rho = \frac{k^2+2}{2-k^2}\omega_0^2$, $p = \omega_0^2$
Sol^3	$h = \lambda\mathrm{e}^{-2z}\mathrm{d}x^2 + \lambda\mathrm{e}^{2z}\mathrm{d}y^2$ $+\frac{1}{\lambda^2}\mathrm{d}z^2$	$f = \frac{2}{\sqrt{k^2-2}}z$, $\|\Omega\|^2 = \frac{4\lambda^2}{k^2-2}$, $k^2 > 2$	$\rho = 2\frac{4-k^2}{k^2-2}\lambda^2$, $p = 0$

Table 6.2: List of Gödel-type solutions with constant vorticity scalar; ω_0 and $R, \lambda > 0$ are constants. The solution with $(N, h) = \mathbb{R} \times H^2$ and $k = 0$ is the classical Gödel spacetime.

Chapter 7

CONCLUSION AND OUTLOOK

The starting point for the investigation presented here is a method for the construction of shear-free spacetimes, with the remaining kinematical quantities specified [GPS+10]. This method is based on the local expression for the metric in comoving coordinates (Eq. (4.1)). By introducing expansion to known causality violating spacetimes, new examples with a different causal structure may be constructed where it is understood that the coordinates are defined globally. However, this procedure considerably lacks control of the curvature, and in general the new Einstein tensor $\text{Ric} - \frac{\kappa}{2} g$ does not correspond to any physically sound matter model.

In this work, we investigate this construction mainly from two angles: Firstly, we require that the shear-free observer field corresponds to a globally defined coordinate vector field. The resulting "tilted" product structure is more general than the original construction which is essentially restricted to spacetimes homeomorphic to $\mathbb{R}^d$. In particular, we can also investigate tilted products with compact fibers and their causal structure.

Secondly, we compute the Ricci and scalar curvature of a large sub-class of shear-free models to obtain formulas that are helpful in solving the Einstein–Hilbert equation for this class of spacetimes. A conformally stationary and a Hubble-isotropic solution modelling a rotating and expanding universe containing a viscous fluid are given.

We conclude with a list of open questions and opportunities for future research.

- When does a spacetime with shear-free, complete reference frame admit a metric splitting into standard form? I.e., what are conditions for such a spacetime to be isometric (or globally conformally equivalent) to a tilted twisted product? In the stably causal case, answers are known for conformally stationary reference frames [JS08], as well as for shear-free, synchronizable reference frames [GRK96, Theorem 3.4].

- What is the geometric structure of more general tilted products, with a base of dimension $d - D > 1$? (See Remark 4.1.5.) In general, the curvature is probably difficult to compute; however, certain additional assumptions on the base might simplify the task.

- The Ricci and scalar curvature formulas for tilted warped products might be useful to study solutions to the gravitational field equations of rigidly rotating matter, which is the case of constant scale parameter.

- Is there a reasonable way to generalize Proposition 6.1.2 relating Sobolev inequalities and causality to the case of non-compact fibers? In fact, using [GT06, Theorem 6.2], if $\xi \in \Lambda^1(N)$ has bounded norm, there is a function f such that $\|\xi - \mathrm{d}f\|_{L^\infty} \leq C\|\mathrm{d}\xi\|_{L^q}$ provided that $B_1(N) = 0$ and the second $L_{\infty,q}$-cohomology class, $H^2_{\infty,q}(N)$, has finite dimension. However, to the best of the author's knowledge, $H^2_{\infty,q}(N)$ is not known for even a single complete noncompact example with $\dim(N) \geq 2$. However, an analogue of Poincaré's lemma is known to hold: $H^k_{\infty,q}(U) = 0$ for $k \geq 1$, whenever $U \subset \mathbb{R}^D$ is a bounded convex domain. Nevertheless, this theorem is not of much use in our context since it only seems to lead to more complicated ways of saying that, locally, (stationary) spacetimes are causally well-behaved.

- We do not know if four-dimensional Gödel-type spacetimes with non-vanishing constant vorticity scalar exist for the case $(N, h) = H^3$. Since Einstein's equations are satisfied for any function f, all that would be left to do is to provide a harmonic function f with a gradient of constant norm. However, to the author's best knowledge no example of such a function defined on hyperbolic 3-space is known. If such a function exists, it cannot be affine [Tas65, Theorem 2]. More precisely, via Bochner's identity it can be shown that if f is such a function, there exists a dual orthonormal basis (E^1, E^2, E^3) with $E^3 = |\mathrm{d}f|^{-1}\mathrm{d}f$, such that the Hessian of f can be written

$$\mathrm{D}\mathrm{d}f = \frac{|\sigma|}{\sqrt{2}}(E^1 \otimes E^1 - E^2 \otimes E^2),$$

where σ is the shear of E_3. Known examples of functions with constant norm of the gradient in hyperbolic space are the Busemann functions [BGS85]. However, Busemann functions are convex and therefore are not harmonic.

BIBLIOGRAPHY

[AC07]   N. Andersson and G. L. Comer, *Relativistic fluid dynamics: Physics for many different scales*, Living Rev. Relativity **10** (2007), no. 1, 1–83.

[Aga84]   V. G. Agakov, *The nonstationary generalization of the Gödel cosmological model*, Gen. Rel. Grav. **16** (1984), 317–323.

[Aub82]   T. Aubin, *Nonlinear analysis on manifolds. Monge–Ampère equations*, Springer, 1982.

[Bau81]   H. Baum, *Spin-Strukturen und Dirac-Operatoren über pseudo-Riemannschen Mannigfaltigkeiten*, Teubner, 1981.

[BB68]   A. Banerjee and S. Banerji, *Stationary distributions of dust and electromagnetic fields in general relativity*, J. Phys. A **1** (1968), no. 2, 188–193.

[BB01]   C. Boubel and L. B. Bergery, *On pseudo-Riemannian manifolds whose Ricci tensor is parallel*, Geometriae Dedicata **86** (2001), 1–18.

[BBS06]   K. Becker, M. Becker, and J. H. Schwarz, *String theory and M-theory: A modern introduction*, Cambridge University Press, 2006.

[Bee76]   J. K. Beem, *Conformal changes and geodesic completeness*, Comm. Math. Phys. **49** (1976), 179–186.

[BEE96]   J. K. Beem, P. E. Ehrlich, and K. L. Easley, *Global Lorentzian geometry*, Marcel Dekker, Inc., 1996.

[Ber03]   M. Berger, *A panoramic view of Riemannian geometry*, Springer, 2003.

[Bes87]   Arthur Besse, *Einstein manifolds*, Springer, 1987.

[BGS85]   W. Ballmann, M. Gromov, and V. Schroeder, *Manifolds of nonpositive curvature*, Birkhäuser, 1985.

[BL04]   H. Baum and F. Leitner, *The twistor equation in Lorentzian spin geometry*, Math. Z. **247** (2004), no. 4, 795–812.

[BN84]   G. S. Birman and K. Nomizu, *The Gauss–Bonnet theorem for 2-dimensional spacetimes*, Michigan Math. J. **31** (1984), no. 1, 77–81.

[BN06]   V. N. Berestovskiĭ and Yu. G. Nikonorov, *Killing vector fields of constant length on Riemannian manifolds*, arXiv:math/0605371v1, 2006.

[BN08]   ―――, *Killing vector fields of constant length on Riemannian manifolds*, Siberian Math. J. **49** (2008), no. 3, 395–407.

[Boc46]   S. Bochner, *Vector fields and Ricci curvature*, Bull. of the Amer. Math. Soc. **52** (1946), 776–797.

[Boh98]   C. Bohle, *Killing and twistor spinors on Lorentzian manifolds*, 1998, Diplomarbeit, Freie Universität Berlin.

[BP03]   K. Behrndt and M. Pössel, *Chronological structure of a Gödel type universe with negative cosmological constant*, Phys. Lett. B **580** (2003), 1–6.

[Bri25]   H. W. Brinkmann, *Einstein spaces which are conformally mapped on each other*, Math. Ann. **94** (1925), 119–145.

[BS05]   A. N. Bernal and M. Sánchez, *Smoothness of time functions and the metric splitting of globally hyperbolic spacetimes*, Comm. Math. Phys. **257** (2005), no. 1, 43–50.

[BSS11]   O. F. Blanco, M. Sánchez, and J. M. M. Senovilla, *Structure of second-order symmetric Lorentzian manifolds*, 2011, http://arxiv.org/abs/1101.5503v2, to appear in: J. Eur. Math. Soc.

[BT82]   R. Bott and L. W. Tu, *Differential forms in algebraic topology*, Springer, 1982.

[Bue96]   P. Bueken, *Three-dimensional Riemannian manifolds with constant principal Ricci curvatures $\rho_1 = \rho_2 \neq \rho_3$*, J. Math. Phys. **37** (1996), no. 8, 4062–4075.

[Car68]   B. Carter, *Global structure of the Kerr family of gravitational fields*, Phys. Rev. **174** (1968), 1559–1571.

[Car00]   S. Carneiro, *A Gödel–Friedmann cosmology?*, Phys. Rev. D **61** (2000), 083506-1–5.

[Car03]   S. M. Carroll, *Spacetime and geometry: An introduction to general relativity*, Prentice Hall, 2003.

[CB09]   Y. Choquet-Bruhat, *General relativity and the Einstein equations*, Oxford University Press, 2009.

[CBDMDB82] Y. Choquet-Bruhat, C. DeWitt-Morette, and M. Dillard-Bleick, *Analysis, manifolds and physics, Part I: Basics*, North-Holland, 1982.

[CF91] N. J. Cornish and N. E. Frankel, *Gravitation in $2 + 1$ dimensions*, Phys. Rev. D **43** (1991), no. 8, 2555–2565.

[Che11] B. Y. Chen, *Pseudo-Riemannian geometry, δ-invariants and applications*, World Scientific, 2011.

[CHL06] A. A. Coley, S. Hervik, and W. C. Lim, *Fluid observers and tilting cosmology*, Class. Quantum Grav. **23** (2006), 3573–3591.

[CJ88] C. J. S. Clarke and P. S. Joshi, *On reflecting spacetimes*, Class. Quantum Grav. **5** (1988), 19–25.

[Col89] A. A. Coley, *On the thermodynamics of cosmological models with heat conduction*, Phys. Lett. A **137** (1989), 235–238.

[Con01] L. Conlon, *Differentiable manifolds*, 2nd ed., Birkhäuser, Boston, 2001.

[COS01] T. Chrobok, Yu. N. Obukhov, and M. Scherfner, *On the construction of shear-free cosmological models*, Mod. Phys. Lett. A **20** (2001), 1321–1325.

[COS02] ――――, *Shear-free rotating inflation*, Phys. Rev. D **66** (2002), 043518.

[CS00] A. M. Candela and M. Sánchez, *Geodesic connectedness in Gödel type space–times*, Diff. Geom. & Appl. **12** (2000), 105–120.

[CSS02] A. M. Candela, A. Salvatore, and M. Sánchez, *Periodic trajectories in Gödel type space–times*, Nonlinear Analysis **51** (2002), 607–631.

[Der12] A. Derdzinski, *Two-jets of conformal fields along their zero sets*, Cent. Eur. J. Math. (2012), 1–12.

[DPS09] A. Dirmeier, M. Plaue, and M. Scherfner, *On conformal vector fields parallel to the observer field*, Adv. Lorentz. Geom. **1** (2009), 21–36.

[dR84] G. de Rham, *Differentiable manifolds*, Springer, 1984.

[DS99] K. L. Duggal and R. Sharma, *Symmetries of spacetimes and Riemannian manifolds*, Kluwer Academic Publishers, 1999.

[Ell67] G. F. R. Ellis, *Dynamics of pressure-free matter in general relativity*, J. Math. Phys. **29** (1967), 1171–1194.

[ER97] A. E. Everett and T. A. Roman, *Superluminal subway: The Krasnikov tube*, Phys. Rev. D **56** (1997), 2100–2108.

[Ev99]   G. F. R. Ellis and H. van Elst, *Cosmological models (Cargèse lectures 1998)*, Theoretical and Observational Cosmology (Marc Lachièze-Rey, ed.), NATO Adv. Study Inst. Ser. C. Math. Phys. Sci., vol. 541, Kluwer Academic Publishers, 1999, pp. 1–113.

[FLGRKÜ01]   M. Fernández-López, E. García-Río, D. N. Kupeli, and B. Ünal, *A curvature condition for a twisted product to be a warped product*, manuscripta math. **106** (2001), 213–217.

[Fol70]   G. B. Folland, *Weyl manifolds*, J. Diff. Geom. **4** (1970), 145–153.

[Fri99]   A. Friedmann, *On the possibility of a world with constant negative curvature of space*, Gen. Rel. Grav. **31** (1999), no. 12, 2001–2008, received 1924.

[FS03]   J. L. Flores and M. Sánchez, *Causality and conjugate points in general plane waves*, Class. Quantum Grav. **20** (2003), 2275–2291.

[FV04]   D.B. Fuchs and O. Ya. Viro, *Topology II: Homotopy and homology. Classical manifolds*, Springer, 2004.

[FW62]   R. W. Fuller and J. A. Wheeler, *Causality and multiply connected space–time*, Phys. Rev. **128** (1962), 919–929.

[Ger71]   R. Geroch, *A method for generating solutions of Einstein's equations*, J. Math. Phys. **12** (1971), no. 6, 918–924.

[GI91]   J. R. Gott III., *Closed timelike curves produced by pairs of moving cosmic strings: Exact solutions*, Phys. Rev. Lett. **66** (1991), 1126–1129.

[GIL98]   J. R. Gott III. and L.-X. Li, *Can the universe create itself?*, Phys. Rev. D **58** (1998), 023501.

[GKS05]   M. Gürses, A. Karasu, and Ö. Sarioglu, *Gödel-type metrics in various dimensions*, Class. Quantum Grav. **22** (2005), 1527–1543.

[GMSW04]   J. P. Gauntlett, D. Martelli, J. Sparks, and D. Waldram, *Sasaki–Einstein metrics on $S^2 \times S^3$*, Adv. Theor. Math. Phys. **8** (2004), 711–734.

[GNS11]   P. Gilkey, S. Nikčević, and U. Simon, *Geometric realizations, curvature decompositions, and Weyl manifolds*, J. Geom. Phys. **61** (2011), no. 1, 270–275.

[Göd49]   K. Gödel, *An example of a new type of cosmological solutions of Einstein's field equations of gravitation*, Rev. Mod. Phys. **21** (1949), 447–450.

[GPS03]   A. García-Parrado and J. M. M. Senovilla, *Causal relationship: A new tool for the causal characterization of Lorentzian manifolds*, Class. Quantum Grav. **20** (2003), 625–664.

[GPS+10]   M. Gürses, M. Plaue, M. Scherfner, T. Schönfeld, and L. A. M. de Sousa jr., *On spacetimes with given kinematical invariants: Construction and examples*, Gödel-type Spacetimes: History and New Developments, Kurt Gödel Society, 2010.

[GPS11]   M. Gürses, M. Plaue, and M. Scherfner, *On a particular type of product manifolds and shear-free cosmological models*, Class. Quantum Grav. **28** (2011), 175009.

[Gra67]   A. Gray, *Pseudo-Riemannian almost product manifolds and submersions*, J. Math. Mech. **16** (1967), no. 7, 715–737.

[GRK96]   E. García-Río and D. N. Kupeli, *Singularity versus splitting theorems for stably causal spacetimes*, Ann. Global. Analysis Geom. **14** (1996), 301–312.

[GT06]   V. Gol'dshtein and Marc Troyanov, *Sobolev inequalities for differential forms and $L_{p,q}$-cohomology*, J. Geom. Analysis **16** (2006), no. 4, 597–631.

[Gür10]   M. Gürses, *Gödel type metrics in three dimensions*, Gen. Rel. Grav. **42** (2010), 1413–1426.

[Hal92]   G. S. Hall, *Weyl manifolds and connections*, J. Math. Phys. **33** (1992), 2633–2638.

[Haw92]   S. W. Hawking, *The chronology protection conjecture*, Phys. Rev. D **46** (1992), 603–611.

[HE73]   S. W. Hawking and G. F. R. Ellis, *The large scale structure of space-time*, Cambridge University Press, 1973.

[HO03]   F. W. Hehl and Yu. N. Obukhov, *Foundations of classical electrodynamics*, Birkhäuser, 2003.

[HP88]   W. Hasse and V. Perlick, *Geometrical and kinematical characterization of parallax-free world models*, J. Math. Phys. **29** (1988), 2064–2068.

[HP99]   ———, *On spacetime models with an isotropic Hubble law*, Class. Quantum Grav. **16** (1999), 2559–2576.

[HR72]   H. Holmann and H. Rummler, *Alternierende Differentialformen*, B.I.-Wissenschaftsverlag, 1972.

[HS81]   H. V. Henderson and S. R. Searle, *On deriving the inverse of a sum of matrices*, SIAM Review **23** (1981), no. 1, 53–60.

[HS91]   G. S. Hall and J. D. Steele, *Conformal vector fields in general relativity*, J. Math. Phys. **32** (1991), 1847–1853.

[Ish01]   C. J. Isham, *Modern differential geometry for physicists*, World Scientific, 2001.

[Isl04]   J. N. Islam, *An introduction to mathematical cosmology*, Cambridge University Press, 2004.

[Jos93]   P. S. Joshi, *Global aspects in gravitation and cosmology*, Clarendon Press, 1993.

[Jos02]   J. Jost, *Riemannian geometry and geometric analysis*, Springer, 2002.

[JS08]   M. A. Javaloyes and M. Sánchez, *A note on the existence of standard splittings for conformally stationary spacetimes*, Class. Quantum Grav. **25** (2008), no. 16, 168001.

[Ker63]   R. P. Kerr, *Gravitational field of a spinning mass as an example of algebraically special metrics*, Phys. Rev. Lett. **11** (1963), 237–238.

[KN96]   S. Kobayashi and K. Nomizu, *Foundations of differential geometry, Volume I*, Wiley, 1996.

[KO93]   V. A. Korotky and Yu. N. Obukhov, *General-relativistic cosmology with expansion and rotation*, Russ. Phys. J. **36** (1993), no. 6, 568–573.

[KR97]   W. Kühnel and H.-B. Rademacher, *Essential conformal fields in pseudo-Riemannian geometry*, J. Math. Sci. Univ. Tokyo **4** (1997), 649–662.

[Kra02]   S. Krasnikov, *No time machines in classical general relativity*, Class. Quantum Grav. **19** (2002), no. 15, 4109.

[Kri90]   M. Kriele, *A generalization of the singularity theorem of Hawking and Penrose to spacetimes with causality violations*, Proc. R. Soc. London A **431** (1990), 451–464.

[KS07]   S. Kar and S. Sengupta, *The Raychaudhuri equations: A brief review*, Pramana **69** (2007), 49–76.

[KSMH79]   D. Kramer, H. Stephani, M. MacCallum, and E. Herlt, *Exact solutions of Einstein's field equations*, Cambridge University Press, 1979.

[KST88]   C. A. Kolassis, N. O. Santos, and D. Tsoubelis, *Energy conditions for an imperfect fluid*, Class. Quantum Grav. **5** (1988), 1329–1338.

[Lan23]   K. Lanczos, *Über eine stationäre Kosmologie im Sinne der Einstein-schen Gravitationstheorie*, Zeitschrift für Physik A: Hadrons and Nuclei **21** (1923), no. 1, 73–110.

[Lee97]   J. M. Lee, *Riemannian manifolds: An introduction to curvature*, Springer, 1997.

[Lee02]   ———, *Introduction to smooth manifolds*, Springer, 2002.

[Lei06]   T. Leistner, *Screen bundles of Lorentzian manifolds and some generalisations of pp-waves*, J. Geom. Phys. **56** (2006), no. 10, 2117–2134.

[Lem31]   G. Lemaître, *A homogeneous universe of constant mass and increasing radius accounting for the radial velocity of extra-galactic nebulae*, Monthly Notices Roy. Astr. Soc. **91** (1931), 483–490.

[LR08]   B. Lew and B. Roukema, *A test of the Poincaré dodecahedral space topology hypothesis with the WMAP CMB data*, A&A **482** (2008), 747–753.

[LWR$^+$03] J.-P. Luminet, J. Weeks, A. Riazuelo, R. Lehoucq, and J.-P. Uzan, *Dodecahedral space topology as an explanation for weak wide-angle temperature correlations in the cosmic microwave background*, Nature **425** (2003), 593–595.

[Mai66]   S. Maitra, *Stationary dust-filled cosmological solution with $\lambda = 0$ and without closed timelike lines*, J. Math. Phys. **7** (1966), 1025–1030.

[Map]   Maple, *http://www.maplesoft.com/*, Waterloo Maple Inc.

[MAT]   MATLAB, *http://www.mathworks.com/*, MathWorks, Inc.

[Mat87]   Yo. Matori, *Totally vicious space–times and reflectivity*, J. Math. Phys. **29** (1987), no. 4, 823–824.

[McM95]   D. J. McManus, *Riemannian three-metrics with degenerate Ricci tensors*, J. Math. Phys. **36** (1995), no. 1, 362–369.

[Mil76]   J. Milnor, *Curvatures of left invariant metrics on Lie groups*, Adv. in Math. **21** (1976), 293–329.

[Min09a]   E. Minguzzi, *Can God find a place in physics? St. Augustine's philosophy meets general relativity*, arXiv:0909.3876v2, 2009.

[Min09b]   ———, *Chronological spacetimes without lightlike lines are stably causal*, Comm. Math. Phys. **288** (2009), 801–819.

[Mon83]   A. Montesinos, *On certain classes of almost product structures*, Michigan Math. J. **30** (1983), 31–36.

[MPL96] P. Musgrave, D. Pollney, and K. Lake, *GRTensorII: A computer algebra system for general relativity*, Fields Institute Comm. **15** (1996), 313–318.

[MS06] E. Minguzzi and M. Sánchez, *The causal hierarchy of spacetimes*, Proc. Geom. of Pseudo-Riemannian Manifolds with Appl. in Phys., 2006.

[Nar85] R. Narasimhan, *Analysis on real and complex manifolds*, Elsevier, 1985.

[Nar01] F. Narita, *Pseudo-Schwarzian tensor of Weyl manifolds*, Kyungpook Math. J. **41** (2001), 155–161.

[Obu90] Yu. N. Obukhov, *Observations in rotating cosmologies*, Gauge Theories of Fundamental Interactions (Proc. XXXII Semester in Stefan Banach Int. Math. Center, Warsaw, Poland, 1988) (R. Raczka and M. Pawlowski, eds.), World Scientific, Singapore, 1990, pp. 343–366.

[Obu92] ———, *Rotation in cosmology*, Gen. Rel. Grav. **24** (1992).

[Obu00] ———, *On physical foundations and observational effects of cosmic rotation*, Colloquium on Cosmic Rotation (M. Scherfner, T. Chrobok, and M. Shefaat, eds.), Wissenschaft und Technik Verlag, Berlin, 2000, pp. 23–96.

[O'N83] Barrett O'Neill, *Semi-Riemannian geometry with applications to relativity*, Academic Press, 1983.

[OS90] J. M. Ollagnier and J.-M. Strelcyn, *On first integrals of linear systems, Frobenius integrability theorem and linear representations of Lie algebras*, Proc. of the Bifurcations of Planar Vector Fields Meeting, Luminy, France, Sept. 1989 (Jean-Pierre Françoise and Robert Roussarie, eds.), Lecture Notes in Mathematics, vol. 1455, Springer, 1990, pp. 243–271.

[Ozs65] I. Ozsváth, *New homogeneous solutions to einstein's field equations with incoherent matter obtained by a spinor technique*, J. Math. Phys. **6** (1965), no. 4, 590–610.

[Pan83] V. F. Panov, *Spontaneous symmetry breaking in cosmological models with rotation*, Theor. and Math. Phys. **74** (1983), no. 3, 316–320.

[Pat96] V. Patrangenaru, *Classifying 3 and 4 dimensional homogeneous Riemannian manifolds by Cartan triples*, Pac. J. Math. **173** (1996), no. 2, 511–532.

[PK06]   J. Plebanski and A. Krasinski, *An introduction to general relativity and cosmology*, Cambridge University Press, 2006.

[PR93]   R. Ponge and H. Reckziegel, *Twisted products in pseudo-Riemannian geometry*, Geometriae Dedicata **48** (1993), no. 1, 15–25.

[PS10]   M. Plaue and M. Scherfner (eds.), *Gödel-type spacetimes: History and new developments*, Collegium Logicum, vol. 10, Kurt Gödel Society, 2010.

[Ray55]   A. K. Raychaudhuri, *Relativistic cosmology I*, Phys. Rev. **98** (1955), 1123–1126.

[Ray79]   ———, *Theoretical cosmology*, Clarendon Press, 1979.

[Ren10]   A. Rendall, *General relativity*, 2010, lecture notes.

[Rob33]   H. P. Robertson, *Relativistic cosmology*, Rev. Mod. Phys. **5** (1933), 62–90.

[RS98]   A. Romero and M. Sánchez, *Bochner's technique on Lorentzian manifolds and infinitesimal symmetries*, Pac. J. Math. **186** (1998), no. 1, 141–148.

[RS99a]   L. Randall and R. Sundrum, *An alternative to compactification*, Phys. Rev. Lett. **83** (1999), no. 23, 4690–4693.

[RS99b]   ———, *A large mass hierarchy from a small extra dimension*, Phys. Rev. Lett. **83** (1999), no. 17, 3370–3373.

[RT83]   M. J. Rebouças and J. Tiomno, *Homogeneity of Riemannian spacetimes of Gödel type*, Phys. Rev. D **28** (1983), no. 6, 1251–1264.

[Sak96]   T. Sakai, *On Riemannian manifolds admitting a function whose gradient is of constant norm*, Kodai Math. J. **19** (1996), 39–51.

[Sán97]   M. Sánchez, *Structure of Lorentzian tori with a Killing vector field*, American Math. Soc. Trans. **349** (1997), no. 3, 1063–1080.

[Sán99]   ———, *On the geometry of generalized Robertson–Walker spacetimes: Curvature and Killing fields*, J. Geom. Phys. **31** (1999), 1–15.

[Sán06]   ———, *On causality and closed geodesics of compact Lorentzian manifolds and static spacetimes*, Diff. Geom. & Appl. **24** (2006), 21–32.

[Sas60]   S. Sasaki, *On differentiable manifolds with certain structures which are closely related to almost contact structure I*, Tohoku Math. J. (2) **12** (1960), no. 3, 459–476.

[Sch00]   M. Scherfner, *Kinematical invariants in Gödel-type models*, Colloquium on Cosmic Rotation (M. Scherfner, T. Chrobok, and M. Shefaat, eds.), Wissenschaft und Technik Verlag, Berlin, 2000, pp. 97–104.

[SO96]   Yo. Soen and A. Ori, *Improved time-machine model*, Phys. Rev. D **54** (1996), 4858–4861.

[SSS98]   J. M. M. Senovilla, C. F. Sopuerta, and P. Szekeres, *Theorems on shearfree perfect fluids with their Newtonian analogues*, Gen. Rel. Grav. **30** (1998), 389–411.

[ST80]   H. Seifert and W. Threlfall, *A textbook of topology*, Academic Press, 1980.

[Ste64]   S. Sternberg, *Lectures on differential geometry*, Prentice Hall, 1964.

[Ste04]   H. Stephani, *Relativity: An introduction to special and general relativity*, Cambridge University Press, 2004.

[Str10]   C. Stromenger, *Sasakian manifolds: Differential forms, curvature and conformal Killing forms*, Ph.D. thesis, University of Cologne, 2010.

[SW77]   R. K. Sachs and H.-H. Wu, *General relativity for mathematicians*, Springer, 1977.

[Tas65]   Y. Tashiro, *Complete Riemannian manifolds and some vector fields*, Trans. AMS **117** (1965), 251–275.

[Thi97]   W. Thirring, *Classical mathematical physics: Dynamical systems and field theories*, Springer, 1997.

[Tip74]   F. J. Tipler, *Rotating cylinders and the possibility of global causality violation*, Phys. Rev. D **9** (1974), 2203–2206.

[Tri92]   F. Tricerri, *Locally homogeneous Riemannian manifolds*, Rend. Sem. Mat. Univ. Pol. Torino **50** (1992), no. 4, 411–426.

[Upo94]   A. M. Upornikov, *Rotating charged fluid without electric field in general relativity*, Class. Quantum Grav. **11** (1994), 2085–2091.

[VBS84]   E. P. V. Vaidya, M. L. Bedran, and M. M. Som, *Comments on the source of Gödel-type metrics*, Prog. Theor. Phys. **72** (1984), 857–859.

[Vis97]   M. Visser, *Lorentzian wormholes: From Einstein to Hawking*, Springer, 1997.

[vS37]   W. J. van Stockum, *The gravitational field of a distribution of particles rotating around an axis of symmetry.*, Proc. Roy. Soc. Edinburgh A **57** (1937), 135.

[Wal37]   A. G. Walker, *On Milne's theory of world-structure*, Proc. London Math. Soc. 2 **42** (1937), 90–127.

[Wal84]   R. M. Wald, *General relativity*, University of Chicago Press, 1984.

[Wei96]   T. Weinstein, *An introduction to Lorentz surfaces*, de Gruyter, 1996.

[Wic54]   G. C. Wick, *Properties of Bethe–Salpeter wave functions*, Phys. Rev. **96** (1954), no. 4, 1124–1134.

[Wil70]   F. W. Wilson, *Some examples of vector fields on the 3-sphere*, Ann. de l'institut Fourier **20** (1970), no. 2, 1–20.

[YB53]   K. Yano and S. Bochner, *Curvature and Betti numbers*, Princeton University Press, 1953.

VITA

Matthias Plaue studied physics in Heidelberg and Berlin with emphasis on mathematical methods, in particular differential geometry and global analysis. Since 2007, he has been working at the TU Berlin in the field of mathematical relativity while being active in a number of other projects including 3D face recognition, human crowd analysis and mathematics education for engineers.